ENVIRONMENTALLY SUSTAINABLE GROWTH

—

STEVEN COHEN

ENVIRONMENTALLY SUSTAINABLE GROWTH

A Pragmatic Approach

COLUMBIA UNIVERSITY PRESS
NEW YORK

Columbia University Press
Publishers Since 1893
New York   Chichester, West Sussex
cup.columbia.edu

Library of Congress Cataloging-in-Publication Data
Names: Cohen, Steven, 1953 September 6– author.
Title: Environmentally sustainable growth : a pragmatic approach /
Steven Cohen.
Description: First Edition. | New York : Columbia University Press, [2023] |
Includes index.
Identifiers: LCCN 2022043184 (print) | LCCN 2022043185 (ebook) |
ISBN 9780231208642 (Hardback) | ISBN 9780231208659
(Trade Paperback) | ISBN 9780231557535 (eBook)
Subjects: LCSH: Nature—Effect of human beings on. |
Sustainable development—Political aspects. | Environmental policy—
Economic aspects. | Infrastructure (Economics)—Environmental aspects. |
Green technology—Technological innovations. | Environmental
degradation—Prevention. | Environmental ethics.
Classification: LCC GF75 .C635 2023 (print) | LCC GF75 (ebook) |
DDC 338.9/27—dc23/eng20230201
LC record available at https://lccn.loc.gov/2022043184
LC ebook record available at https://lccn.loc.gov/2022043185

Cover design: Elliott S. Cairns

To Lily, Noa, and Adi
A new generation born in joy, hope, and promise . . .

CONTENTS

PREFACE

I t is clear to any objective observer that our planet's resources and ecological well-being are under assault by the actions of the over eight billion people living here on Earth. Those of us in the developed world have a far greater environmental impact than people in the developing world, but there is great political pressure in the developing world to increase material wealth. And that political pressure will lead to increased environmental impacts. Our political stability and security depend on the maintenance of material wealth where it exists and economic growth where people are poor. But to achieve those goals, we must learn how to construct a high-throughput economy that does not destroy our planet's ecosystems. That is the major theme of this book: How do we realistically get from the current economic paradigm to one that permits economic growth while protecting the planet's ecosystems?

This book mainly focuses on the United States—not because this can be achieved without the rest of the world, but without our model and leadership, it is unlikely that this transition can happen without a catastrophe taking place first. I also make the case that this transition has already begun. Moreover, we have already demonstrated that we can utilize advances in organizational management and technology to grow economically while protecting the environment. We need to recognize what we know how to do and utilize our experience and brainpower to end poverty while protecting the planet.

To achieve the complete transition to environmental sustainability, we must undertake the following five actions:

1. Research, measure, and *understand* the current state of environmental degradation.
2. Understand the *causes* of environmental degradation.
3. Develop and implement a *strategy* for reducing pollution and growing a renewable resource–based economy.
4. Build public-sector infrastructure to support environmental sustainability.
5. Change the politics, advocacy, and communication of environmental sustainability.

Our current focus on climate change is essential, but it must not be allowed to crowd other critical environmental issues off the limited bandwidth of our political agenda. COVID-19 demonstrates that issues such as contagious viruses and invasive species can disrupt our way of life and demonstrates that climate is not our only sustainability challenge. Since the U.S. Environmental Protection Agency (EPA) was created back in 1970, we have made enormous progress in using regulations and new technologies to reduce our impact on the environment while allowing economic growth to continue. This suggests that we have the capacity to develop methods of production and consumption that are less damaging to the planet. We simply need to up our game and increase the amount of effort that we devote to this task.

Chapter 1 of this book defines environmental sustainability and provides examples of economic practices that have evolved over the past half-century to enable economic production and consumption that minimize damage to environmental systems. Popular and academic literature and the voices of many environmental advocates claim that the only way to protect the environment is to reduce economic consumption. My view is that consumption must change but cannot be reduced. In the developing world, the demand for increased material wealth is high, and its denial would be politically destabilizing. In the developed world, reduced consumption would also be politically unpopular. In this chapter, I discuss changes in consumption and production now underway that can permit a high quality of life while preserving the planet. This includes a discussion of the sharing economy, increased focus on wellness and health, and a discussion of the concept of the circular economy.

In chapter 2, I present an overview of what we know about the current state of the planet. How much of our forest cover has been paved over? How much marine debris is filling our oceans? How many species have become extinct, and what impact have these extinctions had on food chains and ecosystems? This chapter presents an overview of the environmental conditions that we live under and the trends in air quality, water quality, toxic substances, ecosystem well-being, and global warming. I also discuss our need to conduct more research on the following:

- The impact of climate change
- The persistence of plastics and toxics in the ecosphere
- Deforestation and the loss of biodiversity
- Air, water, and soil pollution
- Invasive species and infectious diseases

Our knowledge of the Earth's physical and living environment is growing, but far from complete. Moreover, as our technologies change and new chemicals and production methods are introduced, human impact on the planet continues to grow. This chapter provides a high-level overview of the current state of the planet. These conditions require mitigation to prevent damage in the future and adaptation to current conditions. Chapter 2 discusses what we know and what we need to know and must learn about our planet's conditions. While chapter 3 analyzes the causes of these impacts, some of the discussion of causal factors is included in chapter 2. Chapter 2 also provides an analysis of the *immediate* causes of environmental degradation, while chapter 3 presents the *underlying* causes of these issues.

I find that the discussion of environmental degradation is best presented along with an understanding of its immediate causes. These causes are then best understood with an analysis of the underlying or fundamental causes. There is overlap here, and chapters 2 and 3 are intended to enhance our understanding of the mess we are in, how we got here, and why it will take time to fix the mess.

In chapter 3, I provide an analysis of our need to increase our understanding of the underlying *causes* of environmental degradation. In my view, the causes of environmental degradation include the following:

- Underregulated production technologies
- Mismanaged manufacturing operations, particularly lack of producer responsibility and production processes designed without accounting for environmental impacts
- Political conflict and warfare
- The absence of environmental values and ethics
- The political pressure for rapid economic development
- Underinvestment in environmental protection technologies
- Consumer demand for products that pollute, growing from the seductiveness of our lifestyle
- Ignorance of science and environmental impacts and insufficient research on those impacts

This ideological opposition to science is growing and deeply problematic. People are willing to accept the benefits of science and technology as if they are conjured by magic, but they resist the science that identifies the costs and proposes methods for mitigating those costs. This chapter discusses why the conditions in chapter 2 have emerged. To preserve the planet, we need to know what causes environmental damage so that as our economy evolves, we can reduce our degree of impact.

Once we understand the environmental conditions that we have created and how and why they were created, we then need a realistic, nonideological strategy for reducing pollution and growing a renewable resource–based economy. Chapter 4 presents this environmental sustainability strategy. Key elements of this strategy are providing public incentives for clean production and consumption. Of course, first we need to clearly define a clean economy by developing generally accepted sustainability metrics. Once we know how to define and measure environmental success, we then need to do the following:

- Generate government support through grants and tax credits for research on green technologies such as batteries, solar cells, and automated waste sorting and reuse facilities.

- Develop lease and buyback business models that close the cycle of production and consumption.
- Provide grants and tax incentives for utilities to modernize their grids and build renewable energy generation and storage facilities. Similar resources need to be provided to local governments to invest in advanced waste sorting and mining technologies.

Former New York City mayor Michael Bloomberg and President Joseph Biden have both connected environmental sustainability to economic growth. A sustainability-centered economic strategy must explain in detail how it works and use hard data to demonstrate the jobs created and the equity and environmental justice goals achieved. Chapter 4 discusses the opportunities created by a transition to a green economy. While many governments have set ambitious decarbonization targets, we need to move beyond this rhetoric to a focus on improving current levels of performance. We need to move away from symbols to operational reality. Government regulation should focus on rigorous and audited measurement of environmental sustainability indicators. Once baselines are established, decades-long efforts should be undertaken to improve and measure performance. Tax benefits should be provided for specific levels of environmental improvement. When politically feasible, financial penalties should strategically, when absolutely necessary, be assessed on organizations that allow or do not improve the deterioration of environmental performance.

Then chapter 5 discusses the investments in infrastructure and organizational capacity needed to implement the strategy introduced in chapter 4. There, I summarize the infrastructure needed to implement a circular rather than linear economy and efforts to reduce pollution and increase the use of renewable resources. The strategy outlined in chapter 4 is not wholly theoretical; but, it requires investments in a variety of types on physical and organizational infrastructure. Some of these sustainability practices have already begun. Subsidies for renewable energy and energy efficiency have been in place in California and New York for several years, as have some federal subsidies for electric vehicles and solar panels. Some businesses have developed models where they buy back obsolete products from consumers. Finally, the Biden administration has proposed funding for electric grid modernization, and some states and their public utility commissions are looking to modernize their electric systems as well. This chapter and the conclusion to this book present examples of the actual projects and investments needed to provide greater depth to the strategy presented in chapter 4.

While chapter 5 and the conclusion detail some of the small-scale efforts that are underway to grow a renewable resource–based economy, more needs to be done to build the infrastructure and capacity needed for a circular economy. This is the main focus of chapter 5. In particular, public investment and public-private partnerships will need to generate several trillion dollars for sustainability infrastructure in the United States over the next decade. While the financing is a heavy lift and by no means certain, this chapter details the infrastructure that must be built in the United States to facilitate a transition to an environmentally sustainable economy.

The transition that I am describing will mainly take place in the private sector. However, just as government built or partnered with the private sector to build the infrastructure needed to support the twentieth-century economy, we need a massive infrastructure rebuild to ensure an environmentally sound twenty-first-century economy. We need public sector–funded infrastructure to support environmental sustainability. This begins with energy: solar, wind, geothermal, hydrogeneration, microgrids, batteries, distributed electricity generation, and high-voltage, long-distance distribution. While I am emphasizing the environmental benefits of rebuilt infrastructure, the benefits to our economic competitiveness should also be understood. Our aging transportation and energy systems add costs to our economy and increase the costs incurred by American businesses. In a global economy where other nations are investing in low-cost, renewable energy and a modern transportation system, our businesses will be competing against companies that have the benefit of modern infrastructure while we make do with old and decaying systems.

Next, we need a water system built for a warming planet. We need investment in desalination and the construction of new water filtration and distribution systems. In dry climates, we will need to recycle wastewater. We also need to reimagine our waste system to mine it for resources. This requires investment in waste-to-energy facilities, automatic methods of sorting waste, recycling, waste reduction, and developing advanced anaerobic digestion technologies to take food waste and return it to farms as fertilizer. We need to do the same with sewage treatment: develop and utilize sewage as a resource for growing food.

In addition to energy and waste, we need infrastructure for transportation. This includes traditional roads and bridges, but also mass transit construction, operation, and maintenance. We need public and private electric vehicle charging infrastructure. Some electric vehicles will recharge at the local convenience store as chargers replace gas pumps; others will recharge on city streets at charging stations built into light poles. Most will recharge in the family driveway. We also need research and development of environmentally sound air and high-speed rail travel. It may be that in addition to electronic vehicles, some vehicles might be powered by hydrogen and other nonfossil fuels.

Another key area of infrastructure is communications technology—rural broadband, urban wireless, and advanced cellular communication must be encouraged and subsidized when necessary. These technologies enable the consumption of ideas, entertainment, and social interaction with very low environmental impacts. The modern, sustainable, brain-based economy is built on low-cost communication and information infrastructure. Universal access to this technology is a requirement of an equitable, just opportunity structure.

Finally, as the COVID crisis has taught us, public health institutions must be rebuilt at the local, national, and global levels. We will see additional pandemics if we do not act now. We need:

- Global virus tracking and reporting
- The creation of local contact tracing and isolation organizational capacity, vaccination, testing, and facilities
- Enhanced health condition monitoring and communication
- Research on virus cures and treatments

Chapter 5 provides details and examples of the infrastructure needed for environmental sustainability. Without public investment, the private market will be unable to turn the profits needed to rapidly transition away from nonrenewable resources. But public investment will not appear by magic. As American infrastructure politics continued along its dysfunctional path in the fall of 2021, it became clear that the political and communication strategies of environmental politics needed an overhaul. The negativity and arrogance of some environmental advocates contribute to a cultural and ideological divide that reduces support for environmental protection. The economic self-interest of fossil fuel interests has not been successfully countered by efforts to villainize those interests.

Chapter 6 discusses the need to articulate a positive vision of an environmentally sustainable lifestyle and promote these images through the media, culture, and entertainment. Environmental sustainability advocates need to create and disseminate positive role models and reduce the focus on environmental symbolism and the emphasis on enemies. People that work for fossil fuel companies are not evil. They are simply trying to support their families. We need to persuade and cultivate rather than shame those holding different views. This chapter examines the need to build consensus-based environmental sustainability politics, built on the need to promote health as well as global economic competitiveness.

Chapter 6 also presents my own perspective about the need to reform environmental politics. In my view, the transition to sustainability should focus on sustainability successes rather than on "evil polluters." Advocates should articulate the attraction of wellness and the health benefits of a clean environment rather than the dangers of illness from toxics. Scare tactics get attention and the media loves them, but they are ultimately negative and divisive. To ensure the transition to a renewable resource–based economy, environmental advocates need to build a wide political consensus rather than insist on a single view of the causes of damage to the planet. Urban environmentalists should restore the traditional alliances with people who hunt and fish. We should also provide natural experiences and images to our increasingly urban population. Some natural experiences will be had outside the city, but some should take place in restored urban parks and wetlands.

Chapter 6 introduces environmental values and ethics and describes their relationship to environmental politics and policy. Alliances with religious groups can be used to connect environmental quality to religion and religious institutions. Even if we differ on other issues, everyone likes to breathe clean air, drink safe water, and eat food free of poison.

A realistic transition to environmental sustainability will be a long, gradual process where we change the definition of the political center and anti-pollution policies will be as accepted as policies that seek to prevent political instability and violence. The speed of this transition will resemble the slow pace of change that we experienced while transitioning from cities that traded natural resources to manufacturing cities and from industrial cities to the service-oriented urban places that many of us now call home. The floods, winds, and fires of climate change and the horrific health and economic impacts of COVID have caused many people to question our ability to dominate nature. This makes new environmental politics possible. We need to understand nature better in order to both live within its boundaries and maintain our current way of life. As I often say, "Our species is ingenious and is not suicidal," so I suspect that we will somehow manage the difficult transition to environmental sustainability. In chapter 6, I connect politics and communication to sustainability strategy and management.

The book concludes with a restatement of the path that I believe provides a realistic way to environmental sustainability. It looks to the past for successes that we have had in previous economic transitions and seeks to apply those lessons to our current situation. The COVID and climate crises are fundamentally disrupting the global economy and our way of life. These two crises are elements of a global crisis of environmental sustainability. We are developing and utilizing technologies without considering their impact on the planet. Paradoxically, we will need additional technologies to address the problems caused by current technologies. COVID requires new vaccination and treatment technologies. Climate change requires new energy technologies. Ultimately, a half-century from now, we will need to capture and store some of the excess carbon now trapped in the atmosphere and literally baking in climate change. A failure of collective action and investment will increase the costs of these crises and those to come. In addition to these environmental sustainability crises, we must deal with the more "routine" crises of war and terrorism, from the brutal Russian invasion of Ukraine in 2022 to continued narcoterrorism in Latin America. The drive toward environmental sustainability will need to continue despite the threats posed by nonenvironmental issues.

Environmental degradation could cause food supplies to become infected, fish could fail to spawn, new viruses could harm people and livestock, and the cost of the necessary responses could overwhelm us. This is the moment to develop methods of detecting and acting on new threats and the capacity to develop and implement new technologies at a wide variety of scales: from individual to global. It will be important to change the political dialogue and move beyond culture wars and ideology. The failure to move toward a realistic transition to environmental sustainability will result in a higher-cost transition when the crisis finally hits with full force. The conclusion makes an argument for a less political approach to environmental protection and for a more determined effort to find common ground. New York City mayor Fiorello LaGuardia once famously stated that "there is no Democratic or Republican way to pick up the garbage." We need to bring environmental sustainability out of the culture wars and into the world of basic delivery of government services such as water supply, sewage treatment, and waste management. Our health and well-being are at stake, and these issues are too important to be derailed by ideology or symbolic politics.

The conclusion also makes a central point of the entire book: The transition to environmental sustainability is already underway. Major corporations are internalizing sustainability management. Governments and corporations have begun to reduce their carbon footprint. Sustainability start-ups are bringing new ideas and thinking to the challenge of building a circular, renewable resource–based economy.

A central theme of this book is that the transition to environmental sustainability is in essence a phase of the modernization of our economy. Lifestyle changes are modifying consumption patterns, and our economy is far more dependent on energy, communications, and information technology than before. Our gross domestic product (GDP) is centered around the service economy, and while the communications and information technologies that we depend on are advancing, waste and energy technologies are advancing too slowly, and investment in these technologies could provide massive increases in standards of living. A side effect of this economic modernization would be less pollution from fossil fuels, mining, and waste disposal.

ACKNOWLEDGMENTS

This book seeks to integrate what I've learned as a student of public policy, public administration, management, and environmental policy. It brings together these disparate areas of knowledge to address the question: What is a realistic path to transition from the current economy that is destroying our planet to a future economy that is environmentally sustainable? In addressing that question, I looked at the transitions that I have seen in my lifetime and tried to determine how far along the path to environmental sustainability we have already traveled.

The learning that this book is based on came from many sources, colleagues, and mentors. During my time at SUNY Buffalo, I learned about environmental policy from Professor Lester Milbrath and Professor Richard Tobin. I also learned from my fellow graduate student (and later coauthor) Sheldon Kamieniecki and my close friend and fellow doctoral student Tony Khater. I also learned about management from Professor Marc Tipermas, who continued to teach me at the Environmental Protection Agency (EPA), the consulting firm ICF, and the Willdan Group. Marc was more than a professional mentor—he has been a good friend who taught me about the centrality of family and personal responsibility. I owe Marc more than mere words could ever convey.

At the EPA, I learned from many people who contributed to the lessons conveyed in this book. Among them were Lee Daneker, Bonnie Casper, Mike Cook, Tom Ingersoll, Tom Jorling, Andy Mank, and Ron Brand. I also learned the meaning of dedication to public service. My colleagues at the EPA were determined to make the world a better place and worked tirelessly to do just that.

At Columbia, I learned from many colleagues, including Al Stepan, John Ruggie, David Dinkins, and Jim Caraley. I also learned from Dan O'Flaherty, John Coatsworth, Ester Fuchs, Bob Cook, Jacob Ukeles, Tanya Heikkila, Sara Tjssoem, Mat Palmer, Howard Apsan, Lloyd Kass, Lynnette Widder, Jeffrey Sachs, George Sarrinikolaou, Kelsie de Francia, Alison Miller, Satyajit Bose, Dong Guo, Louise Rosen, Christoph Meinrenken, Mark Cane, Manu Lall, Vijay Modi, Mike Purdy, Robin Bell, and too many others to list here.

One of the main themes of this book concerns the role of the private sector in the transition to a sustainable economy. Over the past decade I have been involved with a company that provides sustainability services called Willdan. A number of people in that firm have taught me about private sector capacity and public-private partnerships including the company's CEO Tom Brisbin and President Mike Bieber, along with my colleagues on the Board of Directors. I wish to thank and acknowledge them here.

For many years, in many locations, I've relied on my good friend and Columbia colleague, Bill Eimicke. Bill has saved me from myself more often than I care to remember, and he has been my partner on scores of writing, consulting, and teaching efforts. The realistic political analysis that I tried to inject in this work is a continuation of a nearly forty-year discussion that he and I have engaged in.

In producing the manuscript itself, I was assisted at all stages by the incomparable and quite brilliant Maya Lugo. I am confident that she will be an important leader in the field of sustainability management as her career advances. She led a team of incredibly talented interns, including Jasmine Chiu, Johnluca Fenton, and Anna Mahowald. In the later stages of developing this book, I was assisted by Sarah Howard, another quite brilliant graduate of Columbia College. She has managed the work of several interns who worked on this book, including Ana Paula Lamas Ovando, Nayan Meshram, Willa Broderick, Eirlys Chui, and Veronica Marotta. I would also like to acknowledge the dedication and professionalism of the people who helped publish this book. Miranda Martin, the wonderful editor at Columbia University Press, and Brian Smith, also of Columbia University Press, who coordinated production. I am also grateful for the thoughtful copyediting of the book which was undertaken by

Susan McClung. Copyediting is a craft and Susan did a wonderful job of clarifying and focusing my manuscript.

While working on this book, I was serving as vice dean of Columbia's School of Professional Studies. Without the excellent team that I work with there, I could never have found the time or bandwidth to write this, so I want to express my gratitude to the dean, Troy Eggers, and my colleagues on the dean's senior staff, Zelon Crawford, Erik Nelson, and (again) Louise Rosen. I also thank the thousands of students I've taught at Columbia. They challenge me, question me, and inspire me. The field that I've chosen changes daily, and to teach these students, I need to study and learn constantly, and that learning process is a great joy, as well as my life's work.

Last but not least, I wish to acknowledge the constant support of my family. My wife, Donna Fishman, has tolerated my obsession with environmental policy and sustainability management since the 1980s. My siblings, Judith, Robby, and Myra, have tolerated me for even longer, although fortunately for them at a greater distance. While they are now gone, I am often guided by the values and lessons taught to me by my parents, Marvin and Shirley. My mind sometimes plays back their views and voices. My daughters, Gabriella and Ariel, and their families, Eitan, Adi, and Noa, and Rob, Becca, and Lily, are the inspiration for this work. I know that I won't be around to see the transition to environmental sustainability, but I hope and even pray that they will.

I've dedicated this book to my three granddaughters—Lily, who was born in the summer of 2017 and lives in Washington Heights, and Noa and Adi, twin girls born on December 26, 2021, in Israel, where they live. My daughters are both wonderful mothers, and I know that whatever family can do to ensure that children thrive, they have done and will always do. They are both professionally accomplished young women, and I love them beyond reason and I'm proud of who they've become and everything that they do.

But families can't stop global warming, contagious viruses, and toxics in the air, land, and water. For that, we need a global village. I wrote this book to make a small contribution to ensuring that my granddaughters will live in a world that is safe from environmental harm. Lily has spent half of her young life masked, in the shadow of COVID. Noa and Adi came into a world suffering from this awful virus, the brutal invasion of Ukraine by Russia, and ongoing threats to peace and security. These will not be the last forms of environmental and human-induced harm that they will confront, but I know that we can do better than we did in the past in preventing future harms.

I am optimistic about the future because I think that nearly all people are essentially good and care about each other. The faith in the future expressed by my daughters in bringing new life into the world inspires me and provides the motivation to devote my own effort to building a better world. To do that, we must break through ignorance, blind ideology, and misplaced priorities. We have a responsibility to those who follow us to do our very best to preserve the Earth, which they will need to survive and thrive.

ENVIRONMENTALLY SUSTAINABLE GROWTH

—

1

DEFINING AND UNDERSTANDING ENVIRONMENTAL SUSTAINABILITY

This chapter provides a definition of environmental sustainability and discusses how our economic practices can evolve to enable economic production and consumption that will minimize damage to environmental systems. As I noted in the preface, consumption must change, but it cannot be reduced. Political stability requires economic growth. According to Ari Aisen and Francisco Jose Veiga:

> Using . . . a sample covering up to 169 countries, and 5-year periods from 1960 to 2004, we find that higher degrees of political instability are associated with lower growth rates of GDP per capita. Regarding the channels of transmission, we find that political instability adversely affects growth by lowering the rates of productivity growth and, to a smaller degree, physical and human capital accumulation. (Aisen and Veiga 2011)

When people cannot feed their children or young people see no future, they gravitate toward extreme politics since they believe that they have nothing to lose. If we stopped economic growth, investor and consumer confidence would be shattered, and along with it the psychological underpinnings of our modern economy. That could easily turn into an economic depression. A worldwide depression would cause more misery than a climate disaster. In this chapter, I discuss changes in consumption and production now underway that can permit a high quality of life while preserving the planet at the same time.

WHAT IS SUSTAINABLE CONSUMPTION AND PRODUCTION?

Goal 12 of the United Nations (UN) Sustainable Development Goals (SDG) is to "ensure sustainable consumption and production patterns." What does that mean? According to the SDG website:

> Worldwide consumption and production—a driving force of the global economy—rest on the use of the natural environment and resources in a way that continues to have destructive impacts on the planet. Economic and social progress over the last century has been accompanied by environmental degradation that is endangering the very systems on which our future development—indeed, our very survival—depends. . . . Should the global population reach 9.6 billion by 2050, the equivalent of almost three planets could be required to provide the natural resources needed to sustain current lifestyles. . . . Sustainable consumption and production is about doing more and better with less. It is also about decoupling economic growth from environmental degradation, increasing resource efficiency, and promoting sustainable lifestyles. ("Goal 12: Ensure Sustainable Consumption and Production Patterns" n.d.)

There are some types of material consumption that are necessary for biological creatures like human beings to survive. Food, water, air, clothing, and shelter are needed for humans to thrive. The shelter can be long lasting and need not be destroyed through consumption. Clothing can wear out and be destroyed, but it is capable of repurposing and reuse. Foods can be reused by using solid waste and sewage as fertilizer, which is then combined with seeds and photosynthesis to grow new foods. It is possible to imagine a system of consumption that does less damage to the environment than our current system does.

One element of a sustainable production and consumption system would be to design consumer products with reuse in mind. Producers would include in their business models mechanisms for retrieving products from consumers. This concept has been defined by the OECD:

> Extended Producer Responsibility (EPR) is a policy approach under which producers are given a significant responsibility—financial and/or

physical—for the treatment or disposal of post-consumer products. Assigning such responsibility could in principle provide incentives to prevent wastes at the source, promote product design for the environment and support the achievement of public recycling and materials management goals. ("Extended Producer Responsibility" n.d.)

EPR could include a deposit that is returned to the consumer at the end of a product's useful life or a bounty that is paid when a product is returned. Another idea is to design products for longer life and avoid the business model of planned obsolescence (Kramer 2012).

In chapter 4, I discuss how to accelerate the transition to a less polluting and renewable, resource-based economy. But the key is to base material production on renewable rather than finite resources. Production processes that reduce waste through the engineering concept of industrial ecology are also important. Here, the idea is to use all material in the production process and avoid emissions, effluent discharges, and solid waste. This requires that engineers find a productive use for all materials and that the production process utilize renewable energy instead of fossil fuels. The goal is to focus attention on the process of production and to work toward eliminating negative environmental impacts. This builds on the waste reduction concepts imbedded in Total Quality Management (W. Edwards Deming, the creator of TQM, in Cohen and Brand 1990).

We are beginning to see signs of production that seek to utilize these principles. The use of automation and artificial intelligence will enable greater precision in production processes (Mohammadi and Minaei 2019). We will see manufacturing occupying less and less of national GDP and labor. This process is well underway. In the United States, 80 percent of GDP is in the service sector (figure 1.1).

Politically, this statistic should translate into greater power for service-based businesses, which are likely to insist on reduction of environmental degradation since these firms do not directly benefit even in the short run from negative environmental impacts. Moreover, a global economy facilitates competition among companies, and the manufacturer that develops the leanest, least wasteful production process will have cost and price advantages over manufacturers that are less efficient.

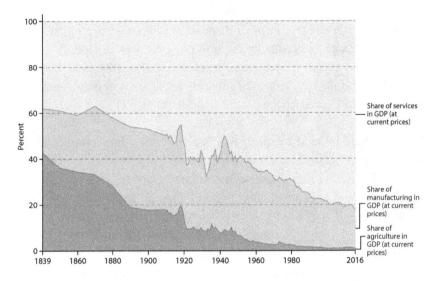

FIGURE 1.1 Shares of GDP by economic sector in the United States from 1839 to 2016.

Source: Lippolis, Nicolas. 2017. "Shares of GDP by Economic Sector, United States, 1839 to 2016." *Our World in Data*, April 24, 2017. https://ourworldindata.org/grapher/shares-of-gdp-by-economic -sector?country=~USA.

WHAT IS A SUSTAINABLE LIFESTYLE?

I have written a great deal about sustainable lifestyles, and my focus is on individual patterns of consumption that have lower environmental impacts than what one might call "unsustainable lifestyles." At its heart, this concept involves a change in material consumption as a means instead of an end. You might aspire to own a home not as an end goal, but rather as a means of engaging in the social and familial activities that the home facilitates. The emphasis is on experience rather than possession. This might lead to using a ride-share service instead of owning a car. It might lead one to rent work or evening clothes from a subscription-based clothing company like Rent the Runway. This is a paradigm shift in the value of consumption and leads to new possibilities in patterns of resource use. The goal is not to eliminate all environmental impact, since that is not possible, but rather to reduce that impact and work continuously at improving our ability to do so.

A discussion of sustainable lifestyles also involves examining how people spend their time. Due to low-cost information and communication, people are spending more time obtaining information and communicating with family and friends than at any time in human history. These activities have very little impact on natural systems. In fact, if you power your smartphone with renewable energy and do not replace the phone frequently, your impact is even lower. Before COVID-19, we saw people moving far from their families but maintaining daily contact by using their smartphones. While some people returned to the nest during the height of the COVID pandemic, that proved to be short term. The changes in family patterns of settlement have led to increased communication and travel. Families that once lived near each other now live far apart. In 2021, the Thanksgiving and Christmas season saw a dramatic revival of holiday travel to prepandemic levels. This is an indication of our pattern of development and employment, which often takes us far from family. Still, our love of family and friends drives us to communicate with them frequently and visit when possible. While travel has proven to be very carbon and resource intensive, airlines and automakers have been working to change technologies, and it is quite possible to imagine a time when their environmental impact might be far lower than today (Griffith 2021).

Another cultural change that facilitates environmental sustainability is the growing focus on wellness. People pay more attention than they once did to diet, exercise, and mental health. Parents focus on the overall wellness of their children, and this attention is a growing element of parenting in the developed world. Once basic necessities are assured, people focus on the positive and negative forces affecting their bodies. This brings to the forefront food, air, and water quality, as well as parks and recreational facilities. Consuming now includes gym memberships, physical training, physical therapy, yoga classes, and other means of self-improvement.

This discussion of sustainable lifestyles is part of my effort to promote a positive vision of environmental sustainability—one that is realistic on a planet of over eight billion people. We can no longer think about a lifestyle that allows us to live off the grid, as one with nature. There are simply too many people and not enough nature for that to happen. So we need to cluster together in places where we can live an exciting lifestyle, exposed to culture, physical activity, nature, entertainment, spiritual engagement, family life, social interaction, and learning, which is fun and fulfilling. That sustainable lifestyle requires sustainable cities reliant on infrastructure and systems that provide us with energy, food, water, and other necessities, while minimizing our negative impact on the planet. The image of parsimony and sacrifice that some environmentalists advocate has little appeal and cannot compete with "the lifestyles of the rich and famous." But a thrilling life full of social engagement, fun, and ideas can. Discussing a new book or film in a café, people and dog watching in the park, biking on a boardwalk, attending a concert, participating in a family or religious celebration, and taking a child to the zoo all seem more interesting to me than showing off my new yacht or hanging out with the Kardashians. Some will always opt for conspicuous consumption, but I think we can build a culture that values friendship, spirituality, family, learning, art, and entertainment over showing off your stuff.

WHAT IS THE SHARING ECONOMY, AND WHAT IS ITS CONNECTION TO SUSTAINABLE CITIES?

Sustainable urban living requires energy-efficient buildings, smart grids, mass transit, and green spaces—but it also seems to be evolving a new approach to owning and using resources. A growing aspect of sustainable urban living is the "sharing economy." Sharing has always been a part of urban life; we have long shared books in public libraries, travel on mass transit, nature in parks, and seats on the stoops of row houses. But in the past few years, cities have seen indications of acceleration in sharing activities and innovations. In cities around the world, people are now welcoming guests into their spare rooms, sharing tools and equipment, and paying for rides in the cars of people they don't know. Start-up businesses are sharing computing space in the cloud and leasing office space and conference rooms by the hour. The sharing economy is growing as some people decide that having access to cars and other resources is more important than owning them. Auto ownership requires insurance, maintenance, and parking—needed expenditures if you drive every day, but less important if you don't. Sharing has been facilitated by the development of the smartphone, the internet, and convenient, inexpensive communication and information.

The size and scale of leading companies operating within the sharing economy, most of which didn't exist a decade ago, now rival some of the world's largest businesses in transportation, hospitality, and other sectors. Cities have many resources that can easily and effectively be redistributed and shared. By allowing people to own less and consume only what they need, fewer resources are wasted, promoting urban sustainability. The sharing economy can lead to more sustainable consumption, while remaining compatible with economic growth.

A growing concern about climate change and environmental sustainability has made the sharing economy an appealing alternative for environmentally conscious consumers. With sharing, less energy is needed for the transportation and production of goods, and less waste is created as everyday products and services are shared among a group of people.

While some cities are embracing the technology, innovation, and new business models that the sharing economy brings, others are resistant to change or have concerns about the safety and quality of these new services, which aren't regulated under the same conditions as traditional services. The biggest challenge to cities is regulatory uncertainty. The growth of the sharing economy has led to political battles in cities all over the world. Governments do not currently have a strong regulatory framework for these new services. Regulating the sharing economy is challenging because existing laws were developed without considering some of the issues presented by a less formal, more dynamic model of production and consumption. These new types of companies do not fit industry regulations perfectly, and they sometimes operate outside the law.

I have written extensively about the sharing economy and do not mean to overstate its importance, but I do think it is an example of more efficient use of resources that can be extended into many areas. What is more important is moving our model of production to consumption away from a linear process that ends with our waste being dumped into a hole in the ground.

THE CIRCULAR ECONOMY

In the December–January 2018 issue of the *European Financial Review*, I wrote:

> In a circular economy, all waste from consumption becomes an input into new production. Inevitably there is some leakage in the tightest circular production process. But the goal is to move from a linear model of production-consumption-waste to one more closely resembling a circular model. I don't think of zero waste as an achievable operational goal, but rather as a model and an aspiration. It is a way to think about resource use and waste management, rather than an absolute target. It requires a paradigm shift or a new way of thinking about consumption and garbage. (Cohen 2017)

There was a time when waste management was a simple matter of digging a hole in the ground and dumping garbage into it. In earlier times, we simply barged municipal garbage out to sea and dropped it into the ocean. In the modern world, there are too many of us, and we produce so much waste that all of it must be reduced, recycled, burned for energy, or somehow treated. Unlike waste in the early part of the twentieth century, some of our waste products are no longer biodegradable and are made of plastics or toxic materials. We are getting better at treating waste, and the available technology is increasingly cost effective and efficient. While we produce more waste than we used to, the rate of waste per capita is going down in the United States, and the amount of waste recycled and treated is going up.

But we can do even better because the technology of waste management is rapidly advancing. It is possible to imagine combining artificial intelligence and automation to build a facility where waste could be automatically sorted and processed for reuse. Even contaminated waste can be treated and then separated for reuse. The goal is to close the system of production and consumption and create a circular economy in which all materials are reused rather than discarded. Such an economy will require more energy to move and treat materials, so as we work to create this economy, we must also work to transition from fossil fuels to renewable energy.

A sustainable, renewable resource–based economy is both possible and necessary. It is also a process that will take decades to complete. A careful, parsimonious approach to the use of physical materials that limits waste will make a production process more efficient and cheaper. We saw this when Total Quality Management reduced manufacturing waste in postwar Japan. In the 1950s and 1960s, the label "made in Japan" signified a low-quality product. By the 1980s, Japanese autos and electronics had developed a reputation for high quality. As technology continues to develop, energy efficiency and renewable energy will beat other forms of energy on price as well. If we maintain a regulatory structure that punishes companies that release poison into the environment, the effort to reduce risk of environmental liability also will result in cleaner production processes. Cities are being retrofitted for sustainability, with sewage treatment and other infrastructure being designed to reduce our environmental impact. Modern economic and social life has become based less on brawn than on brains. That means that potentially more and more attention will be paid to growing our economy while reducing damage to the planet.

The Pace of the Transition to a Circular Economy

During the summer of 2022, I watched the reaction of advocates and experts to the U.S. Supreme Court's decision in *EPA vs. West Virginia*, I was struck then by the dismay that the Environmental Protection Agency (EPA) would no longer be able to implement rapid sweeping changes in the nation's energy system. I have a little news for these experts: The EPA was never going to be able to decarbonize the economy quickly. It was always going to be a slow and gradual process. In Lisa Friedman's *New York Times* article in July 2022 about the EPA's approach to greenhouse regulation in the wake of *EPA vs. West Virginia*, she reported:

> The federal government's piecemeal approach, which is still taking shape, could make it tougher to achieve its goals, many observers said. Power plants that burn fossil fuels are one of the single largest contributors of carbon dioxide to the atmosphere, which is rapidly warming the planet. The Supreme Court's 6–3 decision, which concluded that the E.P.A. lacks broad authority to transform the nation's electrical system away from

fossil fuels . . . did not strip the E.P.A. of its authority to regulate green-house gas emissions, but it allowed only narrower policies to regulate how individual power plants operate. . . . That means the administration's backup strategies are not likely to spur a rapid metamorphosis to clean energy unless the administration acts quickly and aggressively, experts said. (Friedman 2022)

There will be no "rapid metamorphosis." It took Pearl Harbor and the threat of Nazi Germany in World War II to rapidly modernize and con-vert the U.S. economy, and even with FDR's political skill, it was still a close call. The complexity of our energy system and our deep dependence on energy meant that the transition would never be rapid and sweeping. This lack of realism is built on a weak understanding of the pace of social, cultural, organizational, technological, and economic change. While the rate of change isn't quick, the transition to environmental sustainability will happen, and the gradual process of change is already underway.

Economic and technological change has been proceeding since the dawn of civilization, but it has increased in speed over the past two cen-turies. Still, the pace is gradual and often invisible while it is underway. For instance, the transition of the United States from a manufacturing to service economy was not readily visible while it was going on.

Companies that mine minerals and manufacture material goods, like those in the fossil fuel and steel industries, command far less economic power than companies like Google, Amazon, and Microsoft. Ultimately this economic power will translate into political power, some of which will be devoted to ensuring free trade, immigration, and environmental protection, all policies that benefit these companies and their employees. The fossil fuel industry has an interest in burning oil, while Amazon and Walmart need energy but have no reason to prefer fossil fuels to renew-ables. And, based on all those solar arrays on the roofs of Walmarts, I suspect that they've figured out the cost advantages of solar energy.

There are at least six main stages of the transition to a renewable resource–based economy, and some will begin before others are com-pleted. First, the theoretical or conceptual design of the transition must be fully articulated and understood. That phase is now underway. More and more people understand that we can't continue with a linear economy in which every material that we consume produces waste that ends up in a dump. The concept of a circular economy is becoming better known and understood.

The second phase is to attract capital to the green economy. The Securities and Exchange Commission (SEC) climate disclosure rule now being proposed will be fought by some conservatives, but it should prevail since it's the job of the SEC to ensure corporate transparency for investors and more and more investors are asking about environmental risk. Several nongovernmental organizations (NGOs) have developed sustainability metrics, but this effort is too important to leave to the private sector. Governments must ensure that these metrics are developed and audited when applicable. We are already seeing the development of the field of sustainable investment. Over 90 percent of the Standard and Poor's top 500 companies are issuing environment and social governance reports this year, and the size of green investments continues to grow. These are all clear indications that this phase is underway. By the mid-2030s, I suspect this will be a mature element of the world of capital finance.

The third phase is the development of public capital for green infrastructure. The $1.2 trillion infrastructure bill passed in 2021 during the administration of President Joe Biden included over $300 billion for public green investment—the largest single investment in environmental infrastructure in American history. Both California and New York have issued so-called green bonds to provide capital for green public infrastructure investment. This phase will take several decades to build momentum. The antitax, anti-investment ideology of the extreme right has been building momentum from the Ronald Reagan administration through the Tea Party to the Donald Trump administration. These days, we can't even raise the capital needed to repair bridges that are falling down, so these green investments will need to be coupled with user fees whenever possible in order to retire debt. New York City's third water tunnel is an example of green infrastructure funded by water user fees. This tunnel was started half a century ago and completed when Mike Bloomberg was New York City's mayor. All of New York's water was transported from upstate in aging water tunnels, one of which was a century old. New York City increased its water fees to ensure that this critical piece of infrastructure was completed.

The fourth phase is the development of technology that supports a growing economy without damaging the planet. This requires basic research funded by government and tax credits and deductions for corporate research and development. There are a number of technologies that we need to develop. Our current form of waste recycling is woefully inadequate. Home sorting of waste is a good educational exercise, but little more. We need to apply artificial intelligence and robotics to waste sorting and mining. Garbage must be a major resource in the future. We need to turn food waste into fertilizer and mine garbage for plastics, metals, and all forms of material resources. We also need to improve and shrink solar cells with nanotechnology and improve battery capacity to reduce the toxicity of renewable energy technology. We need windows that serve as solar receptors and batteries the size of laptops that cost little and collect solar and wind energy to use during periods without wind and sun. The current solar and battery technology beats fossil fuels in price, but not performance. The price needs to be far lower and solar batteries need to work much better before they drive fossil fuels from the market. But all these technologies are coming.

This phase is well underway, and we are already seeing breakthroughs in new electric vehicle technologies. There are breakthroughs in battery technology. Other improvements have been incremental. Think of the cell phones and laptops that existed at the turn of the twenty-first century and compare them to what we have today. It has been a slow but steady improvement.

The fifth phase is to develop the organizational capacity to utilize new technologies and to seamlessly integrate the behaviors needed to produce goods and services with the least possible impact on the environment. This is well underway. Companies like Land O'Lakes are saving money by deploying agricultural practices that use as little water and chemicals as possible. Drones and satellites, along with robotics and artificial intelligence, are lowering the cost and reducing the environmental impact of farming. Many organizations are building the capacity to minimize their carbon and environmental impacts.

The sixth and final phase is to develop the organizational capacity, infrastructure, and capital needed to ensure widespread implementation, including the retrofitting of older facilities and elements of the built environment. This will be a gradual transition, taking at least a quarter century to complete—although elements of the old economy will persist for many more years.

Environmentalists and experts frequently question the seriousness of policy makers promising to combat climate change. They look for big dramatic gestures and symbols to demonstrate their commitment. The good news is that environmental sustainability is already underway. The bad news is that it will take a long time to reach these goals, and more damage will be caused along the way. We need to focus less on symbols and more on operational reality if we are to truly speed the process and minimize the long-term, irreversible damage to the planet.

Defining the Circular Economy

As noted previously, the first phase is conceptual, and the key concept is the circular economy. There are many people working on and advancing the notion of this type of economy, but none has been as visible and effective as Ellen MacArthur and her foundation. Its website tells her story, and I quote it in full here because I believe that it is both inspirational and a profound example of the power of our planet to motivate:

> In 2005, Ellen MacArthur became the fastest solo sailor to sail around the world. Five years later, she set up the Foundation in her name to accelerate the transition to a circular economy. After circling the globe—carrying everything she needed with her—she returned with new insights into the way the world works, as a place of interlocking cycles and finite resources, where the decisions we make today affect what's left for tomorrow.
>
> Spending 71 days alone at sea, confronted by the awesome power and dazzling beauty of nature, Ellen began to ponder the fragility of the systems we've built. Her boat was her world, and her survival was entirely dependent on the limited food, fuel, and other supplies she'd brought with her. She realised that our global economy is no different—it relies completely on the finite resources we extract, use, and then dispose of.

"No experience in my life could have given me a better understanding of the word finite.*"* [emphasis added]

When she returned, she began a new journey of learning to understand how our economy works. She realised that the solutions to our biggest problems don't just lie in the way we make energy, but also in the way we use materials. Everything we use is in limited supply, from the precious metals in our computers and phones to the sand in cement used to make buildings.

She found that the linear system in which we live is fundamentally flawed. She asked herself what would a successful economy that uses things, rather than uses them up look like? After talking with business leaders, engineers, and other experts, she concluded that building a system that could work in the long term is within our reach. But we would need to transform our extractive, throwaway economic model to one that was based on the principles of a circular economy—an economy designed to keep materials in use, eliminate waste and regenerate natural systems. ("Ellen's Story" 2022)

The transition to a renewable resource–based economy is by necessity the construction of a circular economy. This will be a generation-long process, and it will never be perfect. There will inevitably be leakage. But a vision must be articulated before it can be effectuated, and no one has had a greater impact than Ellen MacArthur in promoting the vision of a circular economy.

While I believe that we will utilize technologies that we do not yet have to achieve a circular economy, a number of other scholars are skeptical about the concept. For example, in a paper in the journal *Ephemera*, Francisco Valenzuela and Steffen Böhm observed:

[Z]ero-waste practices can divert our attention from the planned obso-lescence that has been built into the production and marketing of prod-ucts devised by companies like Apple. . . . Quite simply, as the Apple brand proudly displays its achievements in complying with design-for-recycling standards of production . . . the public turns oblivious of the environmental consequences of Apple's competitive business strategy,

which seeks for consumers to dispose of old versions of Apple products in favour of new releases as quickly as possible. (Valenzuela and
Böhm 2017)

In my view, this critique is of an older version of Apple, which made
most of its money selling iPhones and Macs. Currently, an increased
amount of Apple's profits comes from cloud computing, software, games,
and entertainment. Computer hardware is becoming a commodity,
and software and entertainment are where the largest profits are made.
People are replacing their laptops and other devices less frequently than
they used to and are spending more of their money on applications and
entertainment.

Some believe that a circular economy is not feasible in a growth-based
economy. As observed by Raz Godelnik:

Even if we assume that circular strategies can generate meaningful
reductions in environmental impacts on a micro level (at the level of
the product or business model), a key challenge is whether this can be
done on a macro level, be that a company, an industry or an economy
as a whole. The main debatable variable is growth in economic activity.
For example, if a company reduces the carbon footprint of a product
by 50 percent using circular design strategies, but doubles the sales of
this product, then the total carbon footprint will not change at all. There
seems to be an inevitable tension between the sustainability ambitions
of the circular economy and the growth-based economy it is situated in.
(Godelnik 2021)

Echoing this theme, Kris De Decker, the creator of *Low-tech Magazine*,
argues:

Growth makes a circular economy impossible, even if all raw materials
were recycled and all recycling was 100 percent efficient. The amount of
used material that can be recycled will always be smaller than the material needed for growth. To compensate for that, we have to continuously
extract more resources. (De Decker 2018)

Further, it is possible that the move toward circularity will bolster unsustainable patterns of consumption. Kevin Moss, the global director at the Center for Sustainable Business, stated:

> Consumers still love to buy new clothes. Worldwide, the apparel and footwear market is projected to grow about 5 percent yearly through 2030, resulting in over 100 million tons of production. At least in richer nations, it's easy to imagine a scenario where consumers clean out their closets, donate their clothes to thrift stores and then feel entitled to restock. (Moss 2019)

I think that the concept of the circular economy and the overall issue of environmental sustainability must avoid the trap of trendline thinking. The current patterns of production and consumption will need to be disrupted in order for our economy to move from a linear economy to circularity. I see the possibility of circularity based on the following quite imaginable technological breakthroughs:

- Waste recycling based on an unsorted, mixed-waste stream, separated by automation and artificial intelligence that enable highly efficient mining of resources from waste streams
- Solar energy cells made smaller and more efficient by nanotechnology, and batteries made smaller and more efficient, driving fossil fuels out of the electricity generation process

This would facilitate a renewable energy base for a waste-processing system funded by the mining of raw materials.

ENVIRONMENTAL SUSTAINABILITY AND SUSTAINABILITY MANAGEMENT

My interest in environmental policy began in the fall of 1975, when I wandered into Professor Lester Milbrath's graduate course on environmental policy and politics at SUNY Buffalo. Back then, environmental protection was a fringe issue, far out of the mainstream of policy and governance.

In 1977, I started working at the EPA, staffing a working group on public participation in America's brand-new water pollution programs. The agency was seven years old, and five years earlier, Congress had passed the 1972 Water Act over President Richard Nixon's veto. In 1980 and 1981, I worked in the Superfund toxic waste cleanup program, and in 1985, I worked on the EPA's program to eliminate leaking underground storage tanks. In 1987, I was able to start a fledgling concentration in environmental policy at Columbia's School of International and Public Affairs. In 2002, we started the MPA in Environmental Science and Policy program at Columbia. Until Barack Obama became president, the environment was a small, unimportant part of the political scene. We sat at the kids' table. But then President Obama started discussing climate change with global leaders, and by the time we established our master's program in sustainability management in 2010, I could sense the growing momentum behind the field of environmental sustainability and sustainability management.

My interest in what I came to think of as sustainability management began in 2008 and 2009 as an effort to combine my interest in environmental policy with my interest in effective organizational management. I had always considered these two areas separate, but the more I examined the field of organizational management in the twenty-first century, the more I saw environmental issues becoming central to the field of management. This led to my 2010 book, *Sustainability Management*, as well as the development of Columbia's Master of Science program in Sustainability Management. As we designed the curriculum, we developed an area of management study that was required of all students, which we called "the physical dimensions of sustainability management." This included energy efficiency, renewable energy, waste management, climate science, environmental science, ecology, toxicology, hydrology, green architecture, and other topics that had physical or scientific components that managers needed to understand in addition to finance, organizational management, strategy, marketing, quantitative analysis, financial and performance management, and human resource management.

In the decade-plus since then, we have broadened the field of sustainability management to include issues of diversity, equity, inclusion, and access, as well as corporate governance and community impact, and added courses on forests, public space, the circular economy, corporate sustainability reporting, sustainable fashion, environmental justice, and a variety of other new and fascinating topics. The area that has achieved the most attention has been sustainability finance, developed and led by my colleague Professor Satyajit Bose. We have pioneered a range of courses in this area, including green accounting, energy finance, climate finance and sustainable development, financing the clean energy economy, energy markets and innovation, sustainable investing and economic growth, and impact finance. More innovative courses are being developed every semester, and sustainability finance has become a very attractive area to many of our students.

What has been most interesting is that capital markets have finally figured out that corporations are not immune to environmental risk. Climate-induced drought and extreme weather can disrupt business operations. Toxic materials and invasive species can harm ecosystems, and when contagious viruses are involved, they can bring economies to a screeching halt, as they disrupt supply chains that have turned out to be less durable than we had thought.

We are also learning that just as companies need to pay attention to financial and reputational risk, they must also understand and manage their environmental risks. Corporate concern for reputational risk has grown to include areas such as diversity, equity, and treatment of workers. Corporate performance in these areas is increasingly an object of consumer questioning and consumer choice. Corporate governance, which in the United States has traditionally been a bastion of white male domination, is now subject to government and security market regulation. States like California are trying to require diversity on corporate boards located within their jurisdiction. All of this means that management needs to pay more attention to issues of environmental impact, social impact, and corporate governance.

Cristina Banahan and Gabriel Hasson of ISS Corporate Solutions found a positive association between the gender diversity of a corporate board and the environmental, social, and governance (ESG) rating of that corporation. They noted:

Gender diverse boards offer more comprehensive understanding of key company stakeholders. Women may also provide additional insight into consumer trends and consumer priorities for the companies of the boards they serve. For some industries, this can be particularly important; since women make 70 percent of consumer purchasing choices, consumer-focused corporations with gender-diverse boards may have an advantage in decision-making that is more responsive to their customers. More diverse boards may have greater insight into issues driving key stakeholder behavior, particularly in industries focused on consumer preference. Those issues may include corporate social responsibility and environmental stewardship. In fact, academic research finds that corporate social responsibility affects consumer behavior. Thus, diverse boards may be more comprehensive at identifying and responding to these trends, perhaps helping support superior ESG performance. (Banahan 2018)

The mainstreaming of ESG resembles the growth of financial disclosure and the field of accounting back when the SEC was created and worked to make financial risk more transparent during the post-Depression economic recovery of the 1930s. In the 1920s, stock market risk resembled a crooked casino. FDR and SEC chief Joseph Kennedy (the father of future president John F. Kennedy) changed that, building the modern stock market. Accounting began its climb to professional status during that time. I believe that sustainability management is seeing a similar evolution right now.

In fact, the field of sustainability management has evolved well past my original definition, which focused on the physical dimension of sustainability management. As the field evolves, my interest in all its dimensions remains, but my research emphasis remains what I have come to see as a subfield of sustainability management, environmental sustainability. While I am vitally interested in corporate governance, social justice and diversity, equity, inclusion, and access, the focus of *this* work is environmental sustainability. Although I believe that sustainability management correctly includes all these areas, I also believe that each one should be studied, measured, and understood as distinct concepts and what I have termed "subfields" of the overall field of sustainability management.

ESG accounting is not without its critics. Hans Taparia, clinical associate professor at New York University (NYU), observed:

> At the core of the problem is how ESG ratings, offered by ratings firms such as MSCI and Sustainalytics, are computed. Contrary to what many investors think, most ratings don't have anything to do with actual corporate responsibility as it relates to ESG factors. Instead, what they measure is the degree to which a company's economic value is at risk due to ESG factors. (Taparia 2021)

Of course, it is the financial risk posed by environmental sustainability issues that makes sustainability relevant to corporate decision makers and investors. They are not NGOs pursuing environment for its own sake, but companies interested in making money. This is an example of the planet's well-being and corporate self-interest aligning, and in my view, that is a very good development.

A related trend in the investment world is social impact or social value investment, which has been defined and studied by my colleagues William Eimicke and Howard Buffett in their 2018 work *Social Value Investing*. They note that a growing group of investors who are not only interested in profit, but purpose or positive impact on society. Many public pension funds, such as California's, are mindful of the impact of their portfolio's investment and require that it be invested with purpose in mind.

The focus of this book is the practical transition from an economy that is destroying our planet's ecosystems to one that does not. I explore the degraded state of the planet and the causes of that degradation. Environmental injustice is one of the causes of degradation, and ultimately environmental sustainability cannot be achieved without overall sustainability. Nevertheless, as interconnected as each subfield of sustainability may be, each can be discussed and analyzed distinctly. After examining environmental destruction, I then examine how to build and politically sell an economy that sustains the planet's natural systems. The movement to preserve the planet has benefited from many symbolic acts designed to raise awareness. On the first Earth Day, protesters buried an automobile. Countless acts of symbolism have dramatized these issues, from sit-ins to fossil fuel divestment. I do not discount the value of these acts; I often admire and support them. But *now is the time to discard symbolism for pragmatism.*

Environmental sustainability is a central and mainstream issue. Just as our economic life has transitioned from hunting-gathering to agriculture, from trading to manufacturing, and from manufacturing to a service-driven, brain-based economy, it is now time to transition to a renewable resource–based economy that preserves the planet while utilizing it as the basis for our material well-being. This will be a complex technical, organizational, and political challenge, but I believe it can and will be done. We simply have no choice.

WORKS CITED

Aisen, Ari, and Francisco Jose Veiga. 2011. "How Does Political Instability Affect Economic Growth?" International Monetary Fund. https://www.imf.org/external/pubs/ft/wp/2011/wp1112.pdf.

Banahan, Cristina. 2018. "Across the Board Improvements: Gender Diversity and ESG Performance." *Harvard Law School Forum on Corporate Governance*, September 6, 2018. https://corpgov.law.harvard.edu/2018/09/06/across-the-board-improvements-gender-diversity-and-esg-performance/.

Buffett, Howard, and William Eimicke. 2018. *Social Value Investing.* New York: Columbia University Press.

Cohen, Steven. 2016. "Zero Waste in San Francisco and New York: A Tale of Two Cities." *State of the Planet*, March 28, 2016. https://news.climate.columbia.edu/2016/03/28/zero-waste-in-san-francisco-and-new-york-a-tale-of-two-cities/.

Cohen, Steven. 2017. "Understanding the Sustainable Lifestyle." *European Financial Review*, December–January: 7–9.

Cohen, Steven, and Ronald Brand. 1990. "Total Quality Management in the U.S. Environmental Protection Agency." *Public Productivity & Management Review* 14 (1): 99–114. https://doi.org/10.2307/3380525.

De Decker, Kris. 2018. "How Circular Is the Circular Economy?" *Local Futures*, November 7, 2018. https://www.localfutures.org/how-circular-is-the-circular-economy.

"Ellen's Story." 2022. Ellen MacArthur Foundation. https://ellenmacarthurfoundation.org/about-us/ellens-story.

"Extended Producer Responsibility." n.d. Organisation for Economic Co-operation and Development. Accessed October 31, 2022, from https://www.oecd.org/env/tools-evaluation/extendedproducerresponsibility.htm.

Friedman, Lisa. 2022. "E.P.A. Describes How It Will Regulate Power Plants After Supreme Court Setback." *New York Times*, July 7, 2022. https://www.nytimes.com/2022/07/07/climate/epa-greenhouse-gas-power-plant-regulations.html/.

"Goal 12: Ensure Sustainable Consumption and Production Patterns." n.d. U.N. Sustainable Development Goals. Accessed April 9, 2022, from https://www.un.org/sustainabledevelopment/sustainable-consumption-production/.

Godelnik, Raz. 2021. "The Challenge of Circularity in a Growth-Based Economy." *GreenBiz*, October 1, 2021. https://www.greenbiz.com/article/challenge-circularity-growth-based-economy.

Griffith, Dean. 2021. "Sustainability in Aviation: Significant Efforts Underway to Reduce Greenhouse Gas Emissions." *Jones Day*, November 5, 2021. https://www.jdsupra.com/legalnews/sustainability-in-aviation-significant-3311231.

Kramer, Kem-Laurin. 2012. *User Experience in the Age of Sustainability: A Practitioner's Blueprint*. Waltham, MA: Morgan Kaufmann.

Lippolis, Nicolas. 2017. "Shares of GDP by Economic Sector, United States, 1839 to 2016." *Our World in Data*, April 24, 2017. https://ourworldindata.org/grapher/shares-of-gdp-by-economic-sector?country=~USA.

Mohammadi, Vahid, and Saeid Minaei. 2019. "Artificial Intelligence in the Production Process." In *Engineering Tools in the Beverage Industry*, 27–63. New York: Elsevier. https://doi.org/10.1016/B978-0-12-815258-4.00002-0.

Moss, Kevin. 2019. "Here's What Could Go Wrong with the Circular Economy—and How to Keep It on Track." *World Resources Institute*, August 28, 2019. https://www.wri.org/insights/heres-what-could-go-wrong-circular-economy-and-how-keep-it-track.

Taparia, Hans. 2021. "The World May Be Better off without ESG Investing." *Stanford Social Innovation Review*, July 14, 2021. https://ssir.org/articles/entry/the_world_may_be_better_off_without_esg_investing.

Valenzuela, Francisco, and Steffen Böhm. 2017. "Against Wasted Politics: A Critique of the Circular Economy." *Ephemera* 17 (1): 23–60.

2

THE CURRENT STATE OF
ENVIRONMENTAL DEGRADATION

This chapter discusses the current state of environmental degradation. We know a great deal about our ecosystems, but far less than we need to. Medical research funding devoted to understanding the human body dwarfs the amount of funding spent on environmental research. It's natural that we prioritize our own health over the health of the planet, even though it turns out that our health and the planet's health are interconnected. Specifically, we need to do more research on the following:

- The impact of climate change
- The persistence of plastics and toxics in the ecosphere
- Deforestation and the loss of biodiversity
- Air, water, and soil pollution
- Invasive species and infectious diseases

We do not know enough about our planet's physical and living environments. Many technologies have been introduced over the past century, with new chemical combinations and innovative methods of production and little analysis of the impact of these technologies on people or the planet. Our destructive impact continues to grow, and I am convinced that much of the damage could be avoided with care and consideration. This chapter provides a high-level overview of the current state of the planet, with a focus on the United States. I will discuss, on a very general level, what we know and what we need to know and must learn about our planet's conditions. How much of our forest cover remains? How much marine debris is filling our oceans? How many species have become extinct, and what impact have these extinctions had on food chains and ecosystems? I am not trying to assess the cumulative impact of this destruction or guess if we have passed some real or mythical tipping point. My goal is to provide a general sense of the degree of damage. I am assuming that it is not too late to act.

THE IMPACT OF CLIMATE CHANGE

The natural cycles of floods and extreme weather are being intensified by climate change, and massive disasters are damaging farms in the midwestern and western United States. Nearly a century ago, back when the federal government built civilian infrastructure, the Army Corps of Engineers was in the business of understanding and managing floods. Sometimes extreme weather events overwhelmed human efforts at flood management, but typically the engineered environment and the massive infrastructure worked. Today, it appears that the additional impact of climate change is making extreme weather events more so, and the assumptions under which we built flood control infrastructure must now be reexamined and, in many cases, rebuilt. This is true for Midwest farms, and it is also true for cities like New York, where the sewage system is frequently overwhelmed by rain events. In the American West, the issue is extreme drought and the danger of forest fires. Extreme weather events are more frequent and intense due to the impact of human-induced climate change.

Rising ocean surface temperatures are driving stronger and more intense extreme weather events. In fact, there have been seventeen above-average Atlantic hurricane seasons since 1995, the largest stretch on record, and category 4 and 5 events are becoming increasingly frequent (Buis 2020). Farms, small towns, and large cities throughout the American Midwest are suffering under the impact of massive floods. The way that we settled the land and our assumptions about flood control are significantly out of date. Massive damage has resulted from our development patterns, aging and no longer adequate flood control infrastructure, and extreme weather exacerbated by climate change. While many state and local leaders in the Midwest know that they are in trouble and need to respond,

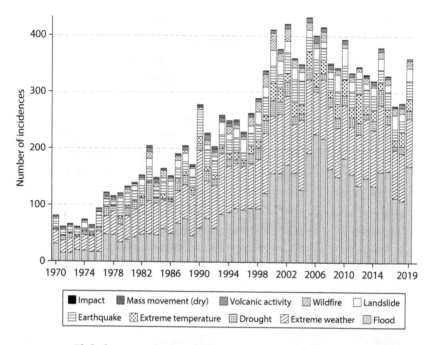

FIGURE 2.1 Global reported natural disasters, organized by type, from 1970 to 2019. This includes both weather and nonweather-related events.

Source: *Our World in Data*. 2020. "Global Reported Natural Disasters by Type, 1970 to 2019," February 3, 2020. https://ourworldindata.org/grapher/natural-disasters-by-type.

some refuse to acknowledge the root cause of the problem. Figure 2.1 graphically presents the increase in disasters during the twentieth and twenty-first centuries.

Even if greenhouse gas pollution ended tomorrow, the current accumulation of these pollutants would guarantee that global warming would continue. Reducing the pace of climate change would help ensure that extreme weather does not become even more extreme, but floods, droughts, fires, and other climate impacts would continue.

When the storm called Hurricane Barry spared New Orleans as it threatened the Gulf Coast back in July 2019, I was again reminded of the growing number of climate-induced extreme weather events hitting the United States. More people live in the pathway of destruction of storms that are wetter, longer lasting, and more intense than ever. While more people are affected by extreme weather events, the U.S. government is not building the institutional capacity to respond to those events. In July 2019, Christopher Flavelle of the *New York Times* observed:

> Three years of crushing natural disasters have left the Federal Emergency Management Agency with even fewer staff available than usual, potentially straining the agency's ability to help victims of Tropical Storm Barry. Less than one-quarter of FEMA's trained disaster workforce of 13,654 people is available to be deployed to Barry or any other emergency, agency documents show, with the rest either already deployed elsewhere or otherwise unavailable. That's down from 34 percent of staff who were available at this point in 2018, and 55 percent two years ago. (Flavelle 2019)

In the fall of 2019, as Hurricane Dorian devastated the Bahamas and headed toward the United States, like many of my neighbors watching, I thought: Could it come to New York City, and are we ready for it? In the years since Superstorm Sandy hit in 2012, the New York metropolitan area had done a great deal to repair its infrastructure and prepare for another storm. Many of the region's beaches were widened, jetties were rebuilt, and storm walls and sand dunes were built and reinforced. Subway tunnels and electric generation plants were made less vulnerable to flooding. New Yorkers understand that climate change is real and they need to do everything they can to get ready for the storms that are coming. But the impact of climate change continues to surprise and often harm us.

Sea level rise and extreme weather pose the same type of existential threat. However, it is not only sea level rise that we need to worry about. In New York City, the remnants of Hurricane Ida in 2021 caused rainfall of five inches per hour in several parts of the city. Ida's rain was so intense that the city experienced a deadly flash flood. This was extreme weather without the impact of sea level rise, but still, the subways became inundated, and people died in basement apartments that were nowhere near the city's 600 miles of coastline (Zaveri, Haag, and Schweber 2021).

While floods from rain and sea level rise are one critical impact of climate change, drought and forest fires are another danger. According to my former colleague, climate scientist Park Williams and his colleagues:

Since the early 1970s, California's annual wildfire extent increased fivefold, punctuated by extremely large and destructive wildfires in 2017 and 2018. This trend was mainly due to an eightfold increase in summertime forest-fire area and was very likely driven by drying of fuels promoted by human-induced warming. Warming effects were also apparent in the fall by enhancing the odds that fuels are dry when strong fall wind events occur. The ability of dry fuels to promote large fires is nonlinear, which has allowed warming to become increasingly impactful. Human-caused warming has already significantly enhanced wildfire activity in California, particularly in the forests of the Sierra Nevada and North Coast, and will likely continue to do so in the coming decades. (Williams et al. 2019, 892)

While scientists continue to connect forest fires to climate change, some conservative ideologues seem to see climate science and climate change policy as a liberal hoax. The issue is far more complex than managing dry leaves and branches in the woods. There are many reasons for the increased damage caused by forest fires in the western United States. These include antiquated electric transmission lines, land use patterns that have brought more people living in and near forests, and a lack of policies that encourage and help people afford the cost of living in more densely settled cities. But clearly, climate change is also a key cause of the increased number, intensity, and impact of forest fires.

Our warming planet is increasing the number and severity of floods, forest fires, hurricanes, and a wide variety of extreme weather events. Climate emergencies are no longer emergencies; they have become routine events. Natural disasters are nothing special and have become a way of life on our warming planet. In late May 2021, the Biden administration doubled the size of a fund that provides local governments with the resources to reduce disaster vulnerability. We need to do far more to prepare for disasters and mitigate their impact.

While some of my colleagues are studying managed retreat from the most climate-vulnerable communities, it is becoming clear that you can run, but you can't hide from climate impacts. The energy grid failure in Texas during February 2021 was a nearly statewide disaster caused by vulnerable infrastructure that was built without consideration of the need for increased resilience. There were few places where Texans could hide from the impact of that extreme weather event.

If we fail to get our act together to build the infrastructure needed to adapt to climate change, the impact on our communities and economy will be devastating. We will be so busy patching potholes, we'll never find the resources to rebuild our roads. Hurricanes will continue to pummel our coasts, forests in the West will continue to burn, and Midwest farms will keep getting flooded. It takes a communitywide effort to build more resilient infrastructure and to rebuild damaged communities. Climate-induced damage is not going to fade with time; instead, we will see growing impacts. In the long run, we will need to reduce greenhouse gases and mitigate climate change, but in the short run, we must invest in the infrastructure needed to adapt to these changes.

THE PERSISTENCE OF PLASTICS AND TOXICS IN THE ECOSPHERE

Plastic pollution is an example of the tragedy of the commons. Plastics are ending up in the oceans because the oceans are not governed by any sovereign nation, so no one takes responsibility for their share of the problem. There is no immediate penalty for dumping plastics, and if we continue along our current path, like the grasslands devoured by farmers' livestock, the oceans will be destroyed. The law of the seas is a noble but inadequate effort to fill the huge gaps in ocean governance. Here in the United States, the management of our waste stream is highly decentralized, and some of the persistent toxics and plastics in the waste stream find their way into groundwater, rivers, and oceans. While some localities do a good job of managing waste, others do not have the resources or interest to do much at all. The fundamental issue of plastic waste must be connected to the overall problem of solid waste, or what most people call "garbage."

Waste production in the United States per capita peaked around 2000, but the growing population means that the volume of waste continues to grow. The volume of waste in rapidly developing nations like India and China is exploding. Increasing amounts of waste in the West and in Japan are recycled or treated in some way, and less garbage is ending up dumped in landfills. Waste-to-energy plants have become more common, as have anaerobic digesters that use food waste to produce fertilizer and natural gas. Plastics are either recycled, burned, or dumped, but when they are dumped, they persist in the environment. Unlike many other forms of waste, they do not biodegrade very easily. At least fourteen million tons of plastic end up in the ocean every year, and this problem is escalating cumulatively year over year: By 2025, there may be as much as one ton of plastic for every three tons of fish. Figure 2.2 presents data on

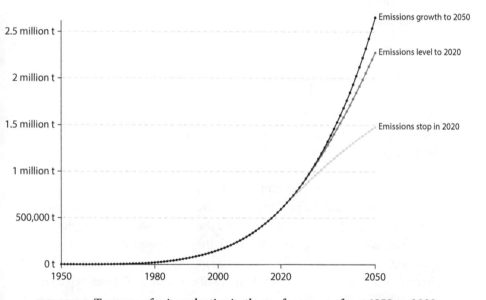

FIGURE 2.2 Tonnage of microplastics in the surface ocean from 1950 to 2020, and extrapolated to 2050 under three emissions scenarios: (1) emissions to the oceans stop in 2020; (2) they stagnate at 2020 emission rates; or (3) continue to grow until 2050 in line with historical plastic production rates.

Source: Our World in Data. 2019. "Microplastics in the Surface Ocean, 1950 to 2050," September 13, 2019. https://ourworldindata.org/grapher/microplastics-in-ocean?time=1950.

the growth of microplastics in the ocean, along with projections of further increases. Microplastics are buoyant plastic materials smaller than 0.5 centimeters in diameter.

According to a recent article in *Science of the Total Environment*:

> Microplastics [MPs] and nano plastics have been observed to cause inflammation, obstruction, and accumulation in organs. . . . MPs have been shown to accumulate in the gills, stomach, and metabolic organs of crabs in in-vitro testing . . . and induce adverse cellular changes in fish. . . . The MPs are persistent in environment and also remain intact inside the living organisms for a long time. Thus, the organisms get prolonged exposure to MPs, which may lead to chronic irritation, resulting in inflammation, cellular proliferation, and necrosis and may compromise immune cells. . . . Furthermore, MPs might act as vectors for various microorganisms. . . . They can release chemicals (organic and inorganic) from their matrixes or adsorb from the environment. (Rahman et al. 2021, 6)

It is not simply plastic waste that leads to toxics in the environment, but also the production of toxic chemicals for use in products. While all new drugs in the United States are reviewed by medical experts in the Centers for Disease Control and Prevention and the Food and Drug Administration before they are authorized for use, new chemicals face no similar rules and are only regulated *after* they are proved dangerous. Many dangerous chemicals continue to be unregulated due to the weakness of the rules and the political power exerted by chemical companies. New drugs are subject to rules based on the precautionary principle, but new chemicals are not. In October 2020, the Biden administration took a small step toward regulating some of the more persistent and dangerous chemicals in wide use. As Lisa Friedman of the *New York Times* reported on October 20, 2021:

> The Biden administration on Monday said it would require chemical manufacturers to test and publicly report the amount of a family of chemicals known as PFAS that is contained in household items like tape, nonstick pans and stain-resistant furniture, the first step toward reducing

their presence in drinking water. Perfluoroalkyl and polyfluoroalkyl compounds, or PFAS refers to more than 4,000 man-made chemicals that are often called "forever chemicals" because they don't break down in the environment. Exposure to the chemicals has been linked to certain cancers, weakened immunity, thyroid disease, and other health effects. (Friedman 2021)

There was disagreement between industry and environmental activists in 2021 about the danger posed by these chemicals, and we heard the usual debate about how much to regulate and what rules should be employed. Friedman's story included a response from the American Chemistry Council, which stated that

about 600 chemicals in the PFAS category are used to manufacture products like solar panels and cellphones, and . . . alternative materials might not be available to replace them. The American Chemistry Council supports the strong, science-based regulation of chemicals, including PFAS substances. But all PFAS are not the same, and they should not all be regulated the same way. (Friedman 2021)

On the other side, as Joseph Winters wrote in *Grist* in October 2020, some environmental advocates saw the approach of the Environmental Protection Agency (EPA) as inadequate to the task. In his piece, he reported that EPA's strategy was to

invest in research to better understand the compounds, restrict PFAS from contaminating the environment, and work to clean up highly polluted areas. By the end of 2021, he said, chemical manufacturers will be required to test and report concentrations for 20 subcategories of PFAS in consumer goods. The EPA will also set enforceable drinking water standards for two of the worst PFAS—known as PFOA and PFOS—and finalize its toxicity assessments for six additional compounds. Some activists lauded the roadmap as a win for environmental justice; PFAS tends to contaminate impoverished neighborhoods and communities of color. . . . But other environmental advocates have said the plan doesn't go far enough. (Winters 2021)

One advocate quoted by Winters derided the EPA's move as a "plan to plan." Others considered the water quality standards inadequate, given the toxicity of some of these chemicals. The EPA's hand in regulating toxics in chemicals has long been weakened by a statutory framework established in the 1970s and 1980s that has not kept up with the exponential increase and vast complexity of new chemical compounds. These chemicals are central to many products in everyday use, and the industry is correct to claim that the application of the precautionary principle to chemical introduction would stifle innovation and impair economic growth. Moreover, if the United States regulated these substances and other nations did not, chemical research and manufacturing would simply move to less regulated nations.

Still, we appear to have made a choice to slowly poison the planet. And while the spread is slow, it has been picking up speed. The chemicals under question are designed to be durable. They do not biodegrade. They are called "forever chemicals" for a reason. Like greenhouse gases, which accumulate in the atmosphere, these chemicals continue to contaminate our ecosystems, water, and air. They are transported through the food chain, ultimately ending up in human bodies. The more of this stuff we make, the more of it remains in our biosphere. The issue is not simply its use in products, but the effluents and emissions created when these durable chemicals are added to a manufacturing process. The chemicals don't only end up on frying pans and in solar cells, but in the water and air as well. It is the very qualities that make these substances useful in products that ensure that they persist in the environment.

In 2018, Veronique Greenwood filed an important report in the *New York Times* on plastic pollution in our coral reefs. She summarized some of the work of the Cornell University professor Joleah Lamb and her colleagues, which estimated plastic pollution in the Asia-Pacific region. This research catalogs plastics on "159 reefs in Australia, Indonesia, Myanmar and Thailand. In a paper released in the journal *Science* . . . they estimate that reefs across the Asia-Pacific region are littered with more than 11 billion pieces of plastic larger than 5 centimeters" (Greenwood 2018).

These plastics are not only ugly, but they attract diseases that can harm reefs, causing them to get sick and die. A key finding from Lamb's study indicates that plastic pollution varies by nation. As Greenwood reported:

> The study shows that it is possible to control the impact of plastic on reefs. Countries that take a great deal of care to keep plastic from entering the ocean—like Australia—see notably lower levels of it on reefs, and the problem was worst in those with poor infrastructure for managing waste, like Indonesia. (Greenwood 2018)

Healthy coral reefs, like other parts of our biosphere, are important for their own qualities, but also for their connection to other forms of life that ultimately create the environment that allows humans to breathe, eat, and exist.

Catrin Einhorn, writing for the *New York Times*, reported:

> While coral reefs cover a tiny fraction of the ocean floor, they provide outsized benefits to people. Their fish supply a critical protein source to hundreds of millions of people. Their limestone branches protect coasts from storms. Their beauty supports billions of dollars in tourism. Collectively, they support an estimated $2.7 trillion per year in goods and services, according to the report, which was issued by the International Coral Reef Initiative, a partnership of countries and organizations that works to protect the world's coral reefs. (Einhorn 2021)

As Barry Commoner once said: "The first law of ecology is that everything is connected to everything else" (Commoner 1971). There is a key lesson in the coral reef story that we see repeated constantly in our economy and way of life. Much of the damage that we do to the planet can be avoided with management, technology, ingenuity, and attention. Some will say that this is an expensive luxury that inhibits economic growth and is particularly problematic in the developing world. The answer is that it is a choice between spending a little bit up front to prevent the problem, or a lot later to clean up the mess. I hate to be trite, but an ounce of prevention really is worth a pound of cure.

Toxic waste is probably the best example of this principle. Dumping toxics in the water or burying barrels underground spreads toxics throughout the environment, and re-collecting those materials costs a lot more than disposing of them safely would have been in the first place. The cleanup of the Hudson River by General Electric (GE) is a dramatic case in point. In 2015, Ted Mann of the *Wall Street Journal* reported that the Hudson River cleanup cost the company $1.7 billion, took seven years, and removed 310,000 pounds of pollutants. Not only did GE have to pay the cost of the cleanup, but according to Mann:

> The river's trustees will make a formal damage assessment that will put a price on what GE owes the state for restoration of resources and wildlife throughout the river ecosystem. Government officials said the company could be asked to pay billions of dollars. (Mann 2015)

There are countless examples of similar expensive cleanups and restoration efforts now underway throughout the United States and in many other nations as well. It is nearly always the case that cleanups are complex and expensive, while pollution control and materials management not only are less expensive, but often stimulate technological innovations that lead to new products and lower-priced production of existing goods and services.

Large-scale environmental disasters such as the British Petroleum (BP) oil spill in the Gulf of Mexico, GE's pollution of the Hudson River, the Fukushima nuclear disaster, and lead getting into the water supply of Flint, Michigan, capture our attention. But most of the costs of environmental mismanagement come from millions of daily decisions to ignore best practices and hope that no one notices. At worst, these sloppy practices are justified as a type of macho management: "If you want to make an omelet, you're going to have to break some eggs." In a planet of eight billion people with countless cameras, drones, and low-cost global communication, we live in an observed world. The practice of "midnight dumping" of toxic waste is no longer simply evil and sloppy, it's stupid—unless your goal is to get caught.

But we can do better, and the coral reef story makes clear how important it is to do so. The goal is to close the system of production and consumption and create a "circular economy," in which all materials are reused rather than discarded. As noted earlier, such an economy will require more energy to move and treat materials, so as we work to create this economy, we must also work to transition from fossil fuels to renewable energy.

Finally, more and more wealth is created in the service economy rather than through manufacturing. In the United States, 80 percent of the GDP is in the service sector. People who own 80 percent of the economy are not likely to allow those who control 20 percent of the economy to poison them or pollute their air and water. Cities are being retrofitted for sustainability with sewage treatment and other infrastructure designed to reduce the environmental impact. Modern economic and social life has become based less on brawn than on brains. That means that more and more attention will be paid to growing our economy while reducing damage to the planet.

But as we improve our current practices, we will need to deal with the awful legacy of the past. Professor Joleah Lamb and her colleagues say that those eleven billion pieces of plastic will soon reach fifteen billion. It will not be easy to address problems like this, but the sooner we get started, the better. We can sustainably manage the environment *and* promote economic growth, but we need to work at it.

DEFORESTATION AND THE LOSS OF BIODIVERSITY

The ecosystems that support life on this planet are a complex web of interconnected and interdependent elements. We do not fully understand all these relationships. Human ignorance, arrogance, and actions have disrupted these interconnected systems since the dawn of civilization. As our technology has advanced, our capacity for inadvertent destruction has exponentially increased. While ancient civilizations might have disrupted only a relatively limited part of the planet, we now have the ability to damage the entire biosphere.

Ecosystem extents and conditions have declined by an average of 47 percent from their natural baselines. Many continue to fall by at least 4 percent each decade. Forests and wetlands have declined by 32 percent and 87 percent, respectively, from preindustrial levels. Live coral cover has been halved as seagrass meadow area and fish stocks plummet. The rate of species extinction, which is still accelerating, has already reached tens to hundreds of times higher than the average rate over the past ten million years (United Nations 2019). Overall, just 3 percent of the world's ecosystems remain intact (Carrington 2021).

The loss of birds and threats to many forms of life are an increasingly accepted part of our modern world. Fires in the Amazon are destroying critical and poorly understood ecosystems, plastics in our ocean are destroying various forms of sea life, lead in our water supply is impairing human health, and toxics in our waste stream are finding their way into our food supply.

As human population has grown, forests have been destroyed, first for cultivation as farmland and then later as sites for housing or commercial enterprises. The services that forest ecosystems provide for absorbing carbon dioxide, filtering and absorbing rainwater, and providing food for forms of life that we rely on are sacrificed when forests are destroyed. When this happens with enough frequency, species become endangered and eventually extinct. We know that monocultures in agriculture increase vulnerability to disease, and we should assume that the loss of biodiversity can endanger life in ways we do not fully understand. The interactive impacts of destruction on interdependent systems add unpredictability to the degree of danger that destruction brings. While we do not know the precise shape of destruction, a report from the United Nations (UN) in 2019 made clear the scope of damage now underway. According to the report:

> Nature is declining globally at rates unprecedented in human history—and the rate of species extinctions is accelerating, with grave impacts on people around the world now likely, warns a landmark . . . report from the Intergovernmental Science-Policy Platform on Biodiversity and Ecosystem Services (IPBES). . . . The Report finds that around 1 million animal and plant species are now threatened with extinction, many within decades, more than ever before in human history. The average abundance of native species in most major land-based habitats has fallen by at least 20 percent, mostly since 1900. More than 40 percent of

amphibian species, almost 33 percent of reef-forming corals and more than a third of all marine mammals are threatened. The picture is less clear for insect species, but available evidence supports a tentative estimate of 10 percent being threatened. At least 680 vertebrate species had been driven to extinction since the 16th century and more than 9 percent of all domesticated breeds of mammals used for food and agriculture had become extinct by 2016, with at least 1,000 more breeds still threatened. (United Nations 2019)

Earth's natural habitats are shrinking rapidly, and perhaps permanently. A recent report from the World Economic Forum suggests:

Climate change and global food demand could drive a startling loss of up to 23 percent of all natural habitat ranges in the next 80 years. . . . Habitat loss could accelerate to a level that brings about rapid extinctions of already vulnerable species. Shrinking ranges for mammals, amphibians and birds already account for an 18 percent loss of previous natural ranges . . . with a jump expected to reach 23 percent by this century's end. . . . Global food demand currently fuels agricultural sectors to increase land use, moving into habitats previously untouched. What results—deforestation—leaves more carbon dioxide in the air, increasing greenhouse gas emissions, the main driver of climate change. In the U.S. alone, agriculture-related emissions measure 11.6 percent of the world's greenhouse gas emissions, which include carbon dioxide, methane and nitrous oxide. (Nairn 2020)

While this degree of destruction raises ethical issues, it is also dangerous because we do not understand its causes or its impact. The unthinking, and at times almost unconscious, destruction remains a phenomenon within the United States despite laws like the National Environmental Policy Act that require environmental impact statements for major projects. It is also a problem globally, with impacts that frequently cross-national boundaries. A World Bank Report in 2021 found that

the collapse of select ecosystem services provided by nature—such as wild pollination, provision of food from marine fisheries and timber from native forests—could result in a decline in global GDP of $2.7 trillion annually by 2030. The Economic Case for Nature underscores the strong reliance of economies on nature, particularly in low-income countries. The report highlights that Sub-Saharan Africa and South Asia would suffer the most relative contraction of real GDP due to a collapse of ecosystem services by 2030: 9.7 percent annually and 6.5 percent, respectively. This is due to a reliance on pollinated crops and, in the case of Sub-Saharan Africa on forest products, as well as a limited ability to switch to other production and consumption options that would be less affected. (Pleming 2021)

AIR, WATER, AND TOXIC SOIL POLLUTION

Climate change, toxic contamination of the land, ecosystem destruction, and air and water pollution are all forms of environmental degradation that we need to learn more about and act to reduce. Each is important. If we decarbonize our economy and stop global warming but destroy our land, air, and water and wipe out the forms of life that we depend on, we may find that some of the damage we've done will be irreversible. But we are learning to apply technology to enable economic growth without increased levels of pollution.

Air Pollution

Air pollution in the United States has declined significantly on nearly every measure since 1980. According to the EPA:

Annual emissions estimates are used as one indicator of the effectiveness of our programs. . . . Between 1980 and 2021, gross domestic product increased 187 percent, vehicle miles traveled increased 111 percent, energy consumption increased 25 percent, and U.S. population grew by 46 percent. During the same time period, total emissions of the six principal air pollutants dropped by 73 percent. (Environmental Protection Agency n.d. "Air Quality")

Toxics in the air have also been reduced, again due to regulation and the development of new technologies. According to the EPA:

From 1990 to 2017 emissions of air toxics declined by 74 percent, largely driven by federal and state implementation of stationary and mobile source regulations. (Environmental Protection Agency n.d. "Air Quality")

This decline is evidence of our ability to make progress and make air pollution less bad, but the work is far from over. The American Lung Association cites the health impact of air pollution as follows:

The "State of the Air" 2021 report finds that despite some nation-wide progress on cleaning up air pollution, more than 40 percent of Americans—over 135 million people—are living in places with unhealthy levels of ozone or particle pollution. The burden of living with unhealthy air is not shared equally. People of color are over three times more likely to be breathing the most polluted air than white people. More than four in ten Americans (41.1 percent—more than 135 million Americans) are living in the 217 counties across the nation with monitors that are capturing unhealthy levels ozone or particle pollution. This is 14.8 million fewer people breathing unhealthy air compared to last year's report, mostly from improved levels of ozone pollution. However, the threat of deadly particulate matter air pollution continues to worsen with each new edition of "State of the Air." This year's report finds an increase of close to 1.1 million people living in areas with unhealthy levels of short-term particle pollution compared to last year's report. (American Lung Association 2021)

Water Pollution

The water pollution story is more complex. During the Donald Trump administration, published reports on water quality appeared to end; and so under President Biden, the EPA had to restart the process of measuring the nation's water resources. Nevertheless, the EPA's Office of Water was able to demonstrate progress in cleaning the nation's waterways over the first half-century of its existence. Drinking water supplies have suffered from decaying infrastructure and lead has been used in water pipes and connectors, but many waterways have seen reductions in pollution. While it is very difficult to study national water quality or the impact of the Federal Water Pollution Control Act of 1972, a half-century after its enactment, Joseph Shapiro of the University of California, Berkeley, and David Keiser of Iowa State took on that mammoth task, publishing a study of U.S. water quality in 2018. In a summary of their study on Berkeley's website, Kara Manke observed:

> The team analyzed data from 50 million water quality measurements collected at 240,000 monitoring sites throughout the U.S. between 1962 and 2001. Most of 25 water pollution measures showed improvement, including an increase in dissolved oxygen concentrations and a decrease in fecal coliform bacteria. The share of rivers safe for fishing increased by 12 percent between 1972 and 2001. (Manke 2018)

An excellent article about the study was published in the *Proceedings of the National Academy of Sciences*. According to its authors, David A. Keiser, Catherine L. Kling, and Joseph S. Shapiro:

> Investments to decrease pollution in rivers, lakes, and other surface waters have constituted one of the largest environmental expenditures in US history. Since 1960, US public and private actors have spent over $1.9 trillion ($2014) to abate surface water pollution. This comes to over $140 per person per year, or over $35 billion total per year. . . . These totals exceed total public and private spending to abate air pollution . . . and they exclude investments to purify drinking water. At peak spending in 1977, these investments represented 0.7 percent of the US gross domestic product (GDP). . . . Expenditures to clean up rivers, lakes, and other surface waters have exceeded the cost of investments to clean up air pollution and also have exceeded the costs of most other US environmental initiatives. Research has found that many of these expenditures have decreased water pollution and has suggested ways to make these investments more effective. . . . A majority of analyses, however, find that these investments'

benefits are less than their costs. . . . This is not the case for most environmental goods, such as air and climate pollution. Are the benefits of these investments truly less than their costs, or are available estimates of costs and benefits biased? We conclude that available estimates of the costs and benefits of water pollution control programs are incomplete and do not conclusively determine the net benefits of surface water quality. (Keiser et al. 2019)

These authors demonstrate the limits of cost-benefit analysis, which they suspect is not measuring the full benefits of sewage treatment. I believe that they are correct. In New York City, we dumped raw sewage into the Hudson River until the North River sewage treatment plant was built in 1984. Since that time, the West Side of Manhattan has been transformed, and the waterfront has evolved from an industrial wasteland into a park with high-end apartments nearby. There is a reason that Riverside Drive is a quarter-mile from the river and there were a hill and train tracks between people's homes and the river. Before the treatment plant was built, the river was an open sewer, and in the summer the odor could be disgusting. Today, you can canoe in the river and ride a bike right next to the water. I doubt very much that the full value of the real estate or the benefit of the new park was part of the cost-benefit analysis of the investment in clean water. Even if it were, I am not certain that its full value could be calculated until long after the new plant was built.

Toxic Soil Pollution

The regulation of toxic substances, toxic waste, and solid waste has been a major element of the EPA's mission since the mid-1970s. In April 2020, the EPA Alumni Association published an excellent report designed as a curriculum guide for high school or college courses, "Waste Management: A Half Century of Progress." According to the report:

Modern societies generate large volumes of waste from industrial, commercial, institutional, and residential activities. In the decades before waste management was federally regulated in the United States, discarded materials contaminated land, groundwater, and waterways and posed increasing risks to public health and the environment. Waste was collected and dumped into unlined landfills—even directly into rivers and the ocean. Municipal landfills were usually located near rivers and streams, allowing liquids and refuse to migrate easily into the water supplies. Waste dump sites were unsanitary, attracting rodents, giving off odors, and creating fire hazards. Trash was often burned in the open, contaminating land, water, and air. Liquids containing flammable and toxic chemicals were discharged into unlined "evaporation" ponds, allowing them to migrate into groundwater and waterways. . . . During the late 1970s and early 1980s, EPA, working with the states, went through a process which transformed the waste industry from an unregulated, thinly capitalized, and often marginal series of unregulated businesses into a well-regulated, updated, and more financially responsible industry. Significant amendments to the SWDA were made with the passage of the Resource Conservation and Recovery Act (RCRA) in 1976. Since this amending legislation is so comprehensive, RCRA is generally referred to without reference to the original 1965 act. Through the combined use of the enforcement and permitting tools provided in RCRA, the waste management industry evolved to become both technically and financially capable of providing the long-term care necessary to successfully manage hazardous waste. (EPA Alumni Association 2020, 3)

This is not to say that waste sites no longer leak and all wastes are contained. But the amount of waste per capita has stopped growing, and the percentage of landfilled waste has been reduced from 89 percent in 1980 to 53 percent in 2017. Figure 2.3 indicates the growth in waste treatment in the United States since 1960.

Due to the Superfund cleanup program, the amount of people exposed to toxic waste has been significantly reduced. The use of technology and regulation has reduced (but in no way eliminated) land pollution in the United States. The amount of waste produced in our linear economy ensures that waste management will remain a major environmental problem. This will continue until we develop a circular economy and close the loop on production from use to end of use and reuse nearly all raw materials, discarding almost none of it.

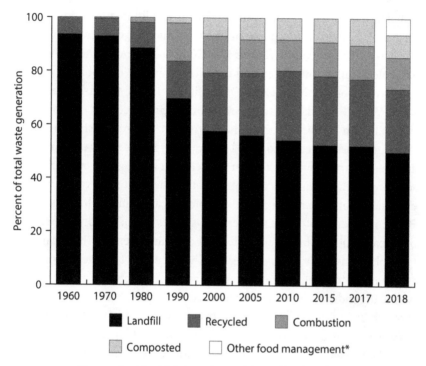

FIGURE 2.3 Percentage of total municipal waste in the United States that is landfilled, recycled, combusted, composted, or otherwise managed.

Source: Center for Sustainable Systems, University of Michigan. 2021. "Municipal Solid Waste Factsheet." Pub. No. CSS04-15. September 2021. https://css.umich.edu/publications/factsheets/material-resources /municipal-solid-waste-factsheet.

Invasive Species and Infectious Diseases

When Christopher Columbus traveled to America, he brought with him weapons that native Americans did not possess, but by far the deadliest weapons that he carried were his crew's European viruses, which the locals could not fight off. Today, we see invasive species in the water supplies of ships. When this water is dumped in the Great Lakes, there is no local predator, and these species grow unchecked. These invasive species unbalance local aquatic ecosystems. This is not limited to water. Due to jet travel, humans are capable of carrying invasive species in the form of viruses. The spread of viruses like COVID-19 around the globe is an example of an unanticipated and unwelcome impact of global travel.

We will not eliminate COVID, and it will not be the last virus of its kind. We will need to learn to tolerate its presence and upgrade our public health infrastructure to keep it under control. Testing, tracing, isolation, masks, vaccination, and treatment are tools that we will need to apply and learn to live with when needed. Just as we go through physical security checks at airports and other venues, we will need to get used to ever-more-sophisticated methods of biological security checks as well. But the presence of these viruses and species is not simply a threat to our health, but to biodiversity itself. Ecosystems tend to be local, and technology makes it relatively easy for living entities to travel beyond the transport systems of the natural environment. This creates a type of economically induced, inadvertent biological warfare. We learned in 2020 through 2022 how difficult it is to overcome a highly contagious virus once it reaches a state of pandemic. The multitrillion-dollar impact of COVID may well result in a global institutional response, but in the short run, the presence and spread of infectious disease remain a threat to environmental sustainability.

CONCLUSION: THE STATE OF THE PLANET EARTH

The natural environment has been modified by humans since we invented agriculture, but the amount of damage done by over eight billion people is now global in scale. The danger is real, and the destruction is widespread. Nevertheless, we have developed rules and technologies that have proved effective in reducing environmental damage while enabling continued economic growth. We have also learned that government can declare land to be parks that must remain natural, and that nongovernmental organizations (NGOs) can purchase land with the same aim. To understand how to mitigate environmental destruction, it is important to understand their fundamental, root causes. That is the subject of chapter 3.

WORKS CITED

American Lung Association. 2021. "State of the Air Report." https://www.lung.org/getmedia /17c6cb6c-8a38-42a7-a3b0-6744011da370/sota-2021.pdf.

Buis, Alan. 2020. "How Climate Change May Be Impacting Storms over Earth's Tropical Oceans." National Aeronautical and Space Administration, March 10, 2020. https:// climate.nasa.gov/ask-nasa-climate/2956/how-climate-change-may-be-impacting -storms-over-earths-tropical-oceans/.

Carrington, Damian. 2021. "Just 3 Percent of World's Ecosystems Remain Intact, Study Suggests." *The Guardian*, April 15, 2021. https://www.theguardian.com/environment /2021/apr/15/just-3-of-worlds-ecosystems-remain-intact-study-suggests.

Center for Sustainable Systems, University of Michigan. 2021. "Municipal Solid Waste Factsheet." Pub. No. CSS04-15. September 2021. https://css.umich.edu/publications /factsheets/material-resources/municipal-solid-waste-factsheet.

Commoner, Barry. 1971. *The Closing Circle: Nature, Man, and Technology*. New York: Knopf.

Einhorn, Catrin. 2021. "Climate Change Is Devastating Coral Reefs Worldwide, Major Report Says." *New York Times*, October 4, 2021. https://www.nytimes.com/2021/10/04 /climate/coral-reefs-climate-change.html.

Environmental Protection Agency. n.d. "Air Quality—National Summary." https://www .epa.gov/air-trends/air-quality-national-summary.

Environmental Protection Agency. n.d. "Protecting the Global Marine Environment." https://www.epa.gov/international-cooperation/protecting-global-marine-environment.

EPA Alumni Association. 2020. "Waste Management: A Half Century of Progress." https:// www.epaalumni.org/hcp/rcra.pdf.

Flavelle, Christopher. 2019. "As Storm Season Begins in the United States, FEMA Is Already Stretched Thin." *New York Times*, July 12, 2019. https://www.nytimes.com/2019 /07/12/climate/barry-storm-fema-staff.html.

Friedman, Lisa. 2021. "A Move to Rein In Cancer-Causing 'Forever Chemicals.'" *New York Times*, October 18, 2021. https://www.nytimes.com/2021/10/18/climate/biden-pfas -forever-chemicals.html?searchResultPosition=2.

Greenwood, Veronique. 2018. "Billions of Plastic Pieces Litter Coral in Asia and Australia." *New York Times*, January 25, 2018. https://www.nytimes.com/2018/01/25 /science/plastic-coral-reefs.html?rref=collection%2Fcolumn%2Ftrilobites.

Keiser, David A., Catherine L. Kling, and Joseph S. Shapiro. 2019. "The Low but Uncertain Measured Benefits of US Water Quality Policy." *Proceedings of the National Academy of Sciences* 116 (12): 5262–5269. https://doi.org/10.1073/pnas.1802870115.

Manke, Kara. 2018. "Clean Water Act Dramatically Cut Pollution in U.S. Waterways." *Berkeley News*, October 8, 2018. https://news.berkeley.edu/2018/10/08/clean-water-act -dramatically-cut-pollution-in-u-s-waterways/.

Mann, Ted. 2015. "GE Nears End of Hudson River Cleanup." *Wall Street Journal*, November 11, 2015. https://www.wsj.com/articles/ge-nears-end-of-hudson-river-cleanup-1447290049.

Nairn, Carly. 2020. "23 Percent of Earth's Natural Habitats Could Be Gone by 2100, Study Finds." World Economic Forum, November 12, 2020. https://www.weforum.org/agenda /2020/11/earth-natural-habitats-destroyed-biodiversity-loss#:~:text=Deforestation%20 and%20climate%20change%20could,to%20die%20sooner%20than%20predicted.

Our World in Data. 2019. "Microplastics in the Surface Ocean, 1950 to 2050," September 13, 2019. https://ourworldindata.org/grapher/microplastics-in-ocean?time=1950.

Our World in Data. 2020. "Global Reported Natural Disasters by Type, 1970 to 2019," February 3, 2020. https://ourworldindata.org/grapher/natural-disasters-by-type.

Pleming, Sue. 2021. "Protecting Nature Could Avert Global Economic Losses of $2.7 Trillion Per Year." World Bank, July 1, 2021. https://www.worldbank.org/en/news /press-release/2021/07/01/protecting-nature-could-avert-global-economic-losses -of-usd2-7-trillion-per-year.

Rahman, Arifur, Atanu Sarkar, Om Prakash Yadav, Gopal Achari, and Jaroslav Slobodnik. 2021. "Potential Human Health Risks Due to Environmental Exposure to Nano- and Microplastics and Knowledge Gaps: A Scoping Review." Science of the Total Environment 757 (February): 143872. https://doi.org/10.1016/j.scitotenv.2020.143872.

United Nations. 2019. "UN Report: Nature's Dangerous Decline 'Unprecedented'; Species Extinction Rates 'Accelerating,'" May 6, 2019. https://www.un.org/sustainabledevelopment /blog/2019/05/nature-decline-unprecedented-report/.

Williams, A. Park, John T. Abatzoglou, Alexander Gershunov, Janin Guzman-Morales, Daniel A. Bishop, Jennifer K. Balch, and Dennis P. Lettenmaier. 2019. "Observed Impacts of Anthropogenic Climate Change on Wildfire in California." Earth's Future 7 (8): 892–910. https://doi.org/10.1029/2019EF001210.

Winters, Joseph. 2021. "EPA Will Finally Regulate 'Forever Chemicals.' Experts Say It's Not Enough." Grist, October 20, 2021. https://grist.org/accountability/epa-will -regulate-forever-chemicals-experts-say-not-enough/.

Zaveri, Mihir, Matthew Haag, and Nate Schweber. 2021. "The Storm's Toll Highlighted New York City's Shadow World of Basement Apartments." New York Times, September 2, 2021. https://www.nytimes.com/2021/09/02/nyregion/nyc-basement -apartments-flooding.html.

3

THE FUNDAMENTAL CAUSES OF ENVIRONMENTAL DEGRADATION

There are many causes of environmental degradation. Most stem from the impact of technology on our lives and on the planet. As part of human development, we first learned how to hunt and gather our food from the world around us, and then we developed technology that allowed us to cultivate food. The ensuing food surpluses allowed our population to grow, and after developing many other new technologies, we eventually established a dominant role on Earth that allowed our species to grow to over eight billion people. In this chapter, I will explore the following fundamental causes of environmental degradation:

- Unregulated production technologies
- Pollution from warfare
- Mismanaged manufacturing operations
- Development of land that should be preserved
- The absence of environmental values and ethics
- The political pressure for rapid economic development and lack of understanding of the connection of environment to economic growth
- Underinvestment in environmental protection technologies
- Consumer demand for products that pollute and our seductive lifestyles
- Ignorance of science and environmental impacts

Some of these causes are due to the ideology of economic growth as a dominant cultural value, some are due to the seductiveness of the material consumption culture that we have created, and some are due to ignorance or incompetence. In this chapter, I will introduce some approaches to addressing specific causes of environmental degradation before turning to a discussion of strategies for transitioning to a less polluting, renewable resource–based economy.

Underregulated Production Technologies

I have long been concerned about the impact of technological complexity on our lives, on representative democracy in the United States, and on other democratic states. The growth of unelected experts in governmental decision-making makes it difficult for elected representatives to influence complex decisions. Linkage from the public to their elected representatives, and then to technical experts, rarely takes place and happens only when issues are urgent. How can people express their policy preferences when they do not understand the technical complexity of the issues they are asked to decide? The process of communicating the impact of new technologies or newly understood problems is far from perfect. In addition, technological complexity is matched by economic complexity and the sheer number of decisions that must be made each day to keep our society and economy functioning. In the current political atmosphere, many trivial symbolic issues reach the political and policy agenda, leaving little room for more important, often complex issues. Climate change, water supply, and disease are ignored while we focus on how or even whether football players should stand during the national anthem.

The goal is to find the optimal level of regulation, one that doesn't stifle innovation or expose people to dangers and risks. Developing and implementing regulations take time, and the half-century-long history of American environmental regulation demonstrates that while the process is slow, it is effective. The slow pace can be frustrating. While the process continues, the environment continues to degrade. Still, the long regulatory process is not entirely dysfunctional because it provides industry with time to adapt and innovate, and allows researchers time to fully understand the impact of toxic substances. Some of the toxic material control policies developed to implement laws enacted in the 1970s were not formally promulgated as regulations until the 1990s. But industry knew that they were coming, substitutes were found, and the economic impact was often slight. Unfortunately, for every toxic chemical that was finally regulated, 1,000 new ones were introduced with unknown levels of environmental impact.

The pace of technological innovation is too rapid for the current regulatory structure to govern. Complaining that the Environmental Protection Agency (EPA) or other federal agencies are not doing enough is attacking the wrong target. The problem is in Congress, and we need a more effective form of regulatory control than the one now in place. I would start by revising our toxic substance laws to require industry to study and report to the public the environmental and health impacts of new chemicals *before* they are used. The EPA now requires chemical producers to study and disclose impacts of select chemicals long *after* the chemicals have been introduced into widespread use. If we tested impacts before use, even if the precautionary principle were not utilized, and companies were free to introduce new chemicals at will, at least the cost of research would be shifted to the private sector, and all new chemicals would be covered, not simply those already thought to be dangerous. Public disclosure would then provide an information base so that the EPA, other regulatory agencies, and advocates could focus their attention on the most dangerous chemicals. The EPA and other federal agencies could add staff to audit industry's environment and health impact studies to reinforce the need for sound science. The current system virtually guarantees continued environmental degradation.

Many decades after toxic substance, hazardous waste, and toxic cleanup legislation were first enacted in the United States, we are long overdue for a major update of those laws. We need to encourage chemical companies to incorporate considerations of environmental impact into their search for new chemical combinations. When energy efficiency standards were added to electrical appliance design, engineers figured out creative ways to manufacture air conditioners and refrigerators so that they used less energy. Let's provide similar incentives for those designing new chemicals. The best way to reduce environmental damage is to prevent it from being designed into new products. We are an ingenious species, and when properly motivated, we can build an economy that doesn't poison people and the planet. Until that creativity is encouraged, we should expect continued environmental destruction.

While I have been focusing on chemicals, there are other elements of the production process that can damage the environment as well. Mining techniques such as strip mining and fracking can damage ecosystems. Other industrial processes can result in toxic releases into the air and water. Companies often place short-term gains over health and safety. From unsafe Boeing 737 Max jets to exploding chemical plants in Houston, we have seen some visible and dramatic impacts of decades of deregulation. This trend picked up momentum under President Ronald Reagan and reached its peak under President Donald Trump. "Regulation" is simply another word for "policing." Cops examine behavior for illegality and when they find it, they turn it over to the courts for adjudication. Rules and their enforcement are requirements of civilization. Without them, we must all protect ourselves in a war of all against all. Deregulation—by definition—leads to increased danger. Defunding regulation gets the same result as defunding the police. We need more effective regulation and policing, not less of it. Effectiveness requires more spending, not less.

The danger of deregulation is that without adequate policing of complex technical processes, the public is left to the mercy of the market. Most businesses are well run and pay attention to safety and emissions. But clearly, some are poorly run and place short-run profits over health and safety. Regulation reinforces correct behavior and justifies investment in safety. On the other hand, deregulation reinforces a Wild West mindset that is inappropriate for the crowded planet that we all share.

Some states are ideologically committed to deregulation. Texas, for example, takes pride in its free market–fueled economic environment, which provides low-cost housing, cheap electricity, and plenty of personal transportation, along with low taxes. The problem is that the absence of rules and governance can come back and create harm if you happen to live on a more crowded and complex planet, like the one we have here on Earth. Climate change, pandemics, traffic jams, toxic air, polluted water, and poisoned land cannot be addressed by the free market alone. In February 2021, a cold snap hit Texas and damaged its electric grid, which was not protected against extreme cold. Its largely unregulated electric utilities had not invested in protection, and its policy of grid isolation from its neighbors left the state without power. As a result, many Texans went without electricity for weeks. Texas's grid troubles were the foreseeable outcome of a reckless effort at deregulation. The *New York Times* reported:

> The crisis could be traced to that other defining Texas trait: independence, both from big government and from the rest of the country. The dominance of the energy industry and the "Republic of Texas" ethos became a devastating liability when energy stopped flowing to millions of Texans who shivered and struggled through a snowstorm that paralyzed much of the state. Part of the responsibility for the near-collapse of the state's electrical grid can be traced to the decision in 1999 to embark on the nation's most extensive experiment in electrical deregulation, handing control of the state's entire electricity delivery system to a market-based patchwork of private generators, transmission companies and energy retailers. . . . With so many cost-conscious utilities competing for budget-shopping consumers, there was little financial incentive to invest in weather protection and maintenance. (Kraus et al. 2021)

POLLUTION FROM WARFARE

Modern wars are not limited to the destruction of military forces by other military forces. The civil war in Syria has involved chemical warfare and mass destruction of ecosystems. The U.S. war in Vietnam resulted in ecological damage from Agent Orange and the use of fire as a military weapon. The presence of nuclear weapons and nuclear power plants creates other risks to ecosystems. In addition to the wanton destruction of Ukraine by conventional weapons, the Russian attack on nuclear sites in 2022 created damage and potentially catastrophic risks to the environment. Ironically, many of my colleagues have long maintained that the best solution to climate change is *nuclear power*. Free of greenhouse gases, nuclear is a powerful, scalable energy technology. We know how to build these plants, and most have operated without incident for many decades. However, in my view, nuclear power is too risky to be a viable option.

While I am far from anti-technology, I have always had two problems with nuclear energy. First, the waste remains toxic for hundreds of thousands of years; and second, there is the possibility of radioactive release due to poor design, poor operation and maintenance, or a terrorist attack. I had never thought that the military of a legitimate sovereign state would be demented enough to attack an operating nuclear power plant, but Vladimir Putin proved me wrong in 2022. His twisted, evil attack on the people of Ukraine has become an attack on the ecological well-being of all of Europe and a large chunk of his own country to boot. Any species that can produce a Putin and give him an army cannot be trusted with the management of such a complex and potentially dangerous technology.

A National Public Radio (NPR) report in 2022 provided graphic detail about the Russian attack on the Zaporizhzhia nuclear power plant in Ukraine. According to the report:

> Last week's assault by Russian forces on the Zaporizhzhia Nuclear Power Plant was far more dangerous than initial assessments suggested, according to an analysis by NPR of video and photographs of the attack and its aftermath. A thorough review of a four-hour, 21-minute security camera video of the attack reveals that Russian forces repeatedly fired heavy weapons in the direction of the plant's massive reactor buildings, which housed dangerous nuclear fuel. Photos show that an administrative building directly in front of the reactor complex was shredded by Russian fire. And a video from inside the plant shows damage and a possible Russian shell that landed less than 250 feet from the Unit 2 reactor building. (Brumfiel et al. 2022)

As if attacking a functioning plant were not sufficient, Russia took over the site of the no-longer-operating Chernobyl nuclear power plant for over a month. The Chernobyl nuclear power plant accident in 1986 spread radioactive materials throughout Europe and was one of the largest nuclear catastrophes in history. As reported by Tucker Reals of CBS News:

Russian forces quickly seized the Chernobyl site after launching their invasion on February 24. Ukrainian officials have said the team of plant operators who ensure safe operations at the decommissioned facility have tried to continue carrying out their work, but under the orders of Russian troops and without being allowed to leave the compound at all. . . . Asked on Thursday about concerns over safety at Chernobyl, U.S. Director of National Intelligence Avril Haines told the Senate Intelligence Committee that the U.S. "should be concerned, but we haven't yet seen anything that takes us from concerned to 'it's a complete crisis.'" Matt Kroenig, who worked on both nuclear and Russia-related issues under the Bush, Obama and Trump administrations, told CBS News senior investigative correspondent Catherine Herridge this week that Putin was weaponing Ukraine's civilian nuclear facilities as part of a strategy to terrorize, and potentially to stage a major nuclear event. (Reals 2022)

Terrorizing civilians and threatening the ecological well-being of the planet constitute an unusual but sadly not unique strategy of homicidal maniacs serving as national leaders. We have seen it for many years with Bashar al-Assad in Syria. We have seen it in Russia beginning in 2022. We saw it in Vietnam when Lyndon Johnson and Richard Nixon served as American presidents. I do not believe that this is the last instance we will see of such lunacy.

Until we develop a form of nuclear power that does not produce dangerous waste and cannot be weaponized as we saw in Ukraine, we should limit the use of this technology as much as possible. The origins of civilian nuclear energy in the "Atoms for Peace" initiative of the Dwight Eisenhower administration was a contrived effort to change the image of nuclear technology from the terror of Hiroshima and Nagasaki to something more benign: electricity that would be "too cheap to meter." Nuclear energy proved to be a little more expensive than that. Nuclear technology was not ready for prime time in the 1950s, and the events in Ukraine in 2022 indicate that it is too dangerous today.

I do not underestimate the threat of climate change, but if I had to choose between a radioactive planet and a warm planet, I'd go with warmth. Fortunately, we do not need to choose. Improved energy efficiency, as well as a generation-long transition to renewable energy, are achievable. The technological breakthroughs needed to decarbonize have started, and I believe that they will pick up momentum. No technology will be free of environmental impact, but I prefer impacts that don't last hundreds of thousands of years and are the type that we can mitigate and reduce as technology develops.

We are highly dependent on energy for many aspects of modern life. Our food, homes, water supplies, waste management systems, and transportation systems all require energy. These human systems can coexist with the natural world if we pay attention to the damage they cause and work to minimize that damage. The problem with nuclear power is that while the probability of damage is low, if damage does occur, its impact can be massive. The lands surrounding Chernobyl and Fukushima have been damaged beyond cost-effective repair. Russia's evil and relentless attack on Ukraine has posed a significant threat via radioactive contamination. According to the World Nuclear Association: "Ukraine is heavily dependent on nuclear energy—it has 15 reactors generating about half of its electricity" (World Nuclear Association 2022). Ukraine's nuclear power plants are distributed throughout that nation, and many have been vulnerable to Russian attack in 2022.

As we watched Russia bomb Ukrainian cities in 2022, and their residents were forced to live without water, food, medicine, and energy, we saw the fragility and interconnectedness of the technological systems that are central to our modern way of life. We observed undeniable courage shown by ordinary people who became extraordinary. They survived without the comforts that we take for granted. But there are limits to what people can endure. Some of these limits are physical, some are psychological, and others come from the sheer toxicity and deadliness of the threats that are posed. Bombs, bullets, and missiles can end life suddenly. Nuclear contamination can present itself suddenly or gradually over time. The technology that we depend on needs to be as resilient, reliable, and safe as possible. Nuclear power fails all those tests, as the war in this nuclear energy–dependent nation has demonstrated. We have learned many difficult lessons from the attack on Ukraine, and the vulnerability of nuclear power to military attack is one of those.

Another is that any military conflict is a fundamental cause of environmental degradation. In addition to nuclear contamination, conventional warfare can have devastating results. In a report filed in mid-April 2022, *New York Times* reporter Emily Anthes observed:

> Since Russian forces invaded Ukraine in February, the world's attention has been focused on the nation's heavily shelled cities. But Ukraine, in an ecological transition zone, is also home to vibrant wetlands and forests and a large swath of virgin steppe. Russian troops have already entered, or conducted military operations in, more than one-third of the nation's protected natural areas. . . . Reports from the ground, and research on previous armed conflicts, suggest that the ecological impact of the conflict could be profound. Wars destroy habitats, kill wildlife, generate pollution and remake ecosystems entirely, with consequences that ripple through the decades. (Anthes 2022)

The delicate balance of ecological well-being and the maintenance of biodiversity are difficult to maintain when human activities are intended to be constructive, but when they are deliberately destructive, the impact on the planet can be massive. While restoration can sometimes follow warfare, some damage, such as species extinction, may well be irreversible.

MISMANAGED MANUFACTURING OPERATIONS

Some environmental damage is caused by organizations that do not pay attention to the impact of their production processes on the environment or do not consider these impacts to be important. In other cases, traditional management processes prove incapable of responding to the impact of climate change and extreme weather on their operation.

After a typhoon hit Fukushima, Japan, in 2011, it left radioactive contamination in its wake. In 2017, it was Houston's petrochemical industry that caught fire, flooded, and leaked toxic chemicals into that region's streets and waterways. And in 2018, environmental poisons came from

North Carolina's pig farms and coal-burning power plants. As Glenn Thrush and Kendra Pierre-Louis reported in the *New York Times* in 2018:

> With floodwaters continuing to rise in the wake of Hurricane Florence, state officials and environmentalists are closely monitoring the breach of a dam that has flooded a hazardous stockpile of coal ash, some of which has spilled into the Cape Fear River. On Friday, Duke Energy shut down a power plant near Wilmington after a dam breach between 100 and 200 feet wide, at the south end of Sutton Lake, allowed floodwaters to swamp two basins containing huge stockpiles of arsenic-laced ash. . . . Coal ash is not the only pollutant to cause North Carolina woes in the wake of Florence. The state is home to 9.7 million pigs that produce 10 billion gallons of manure each year. Most of that manure is stored in large earthen lagoons. In the wake of Florence and the record-breaking amount of water the storm has poured onto the region, a growing number of those lagoons are flooding. (Thrush and Pierre-Louis 2018)

In a world where climate change is increasing the number and intensity of storms and more people and industry are in the pathway of the impact of these storms, we need to rethink the standards that we set for toxic materials and waste storage. We should start to assume that floods, high winds, forest fires, and earthquakes will occur in the places that house toxic materials.

According to a 2019 report from the U.S. Government Accountability Office (GAO):

> Available federal data—from the Environmental Protection Agency (EPA), Federal Emergency Management Agency, National Oceanic and Atmospheric Administration, and U.S. Forest Service—on flooding, storm surge, wildfires, and sea level rise suggest that about 60 percent of all nonfederal National Priorities List (NPL) sites are located in areas that may be impacted by these potential climate change effects. (GAO 2019)

We need to build our toxic storage and processing facilities to withstand the impact of extreme weather events. We also need to detoxify our technology to the extent we can. Nuclear contamination and the toxics from burned coal can be eliminated by the use of renewable energy. We have many reasons to transition to clean energy, and this is yet another argument for renewables. Assuming (as I do) that people will not be giving up their bacon anytime soon, pig waste needs to be reprocessed and detoxified, not left floating in ponds and lagoons.

We've already witnessed the potential impact of climate change on the proliferation of contaminants. According to Hiroko Tabuchi of the *New York Times*:

> A fertilizer plant battered by Hurricane Ida belched highly toxic anhydrous ammonia into the air. Two damaged gas pipelines leaked isobutane and propylene, flammable chemicals that are hazardous to human health. And a plastic plant that lost power in the storm's aftermath is emitting ethylene dichloride, yet another toxic substance. (Tabuchi 2021)

Extreme weather events should challenge traditional notions of effective hazardous materials management. Well established forms of material and waste containment may not work anymore. This means that we need to deploy brainpower and financial resources to ensure that our toxic waste management system is tougher and more resilient than the current system. The financial costs of cleaning up toxics that escape inadequate facilities will be much higher than the costs of building stronger waste management facilities in the first place. Moreover, the health and ecological impacts of failing to remediate the release of toxic substances will cause both physical and financial losses. Toxics are equal opportunity poisons. They harm both rich and poor, and they don't care about anyone's race, ethnicity, or immigration status.

While toxics harm everyone, wealthy people are better able to protect themselves from their impact. They have better health care and are able to evacuate when extreme weather hits. The impacts of toxic chemicals, therefore, are not distributed evenly. The United Nations Human Rights Office of the High Commissioner states:

> The most vulnerable, marginalized and susceptible to toxic exposure face disproportionate threats to life, health and bodily integrity. They

suffer from cancers, miscarriages, birth defects, heart and lung impairments, learning disabilities and diabetes, among others. Those who are most affected are often the most vulnerable members of society. They are people living in poverty, workers, children, minority groups, indigenous peoples, migrants, among other vulnerable or susceptible groups, with highly gendered impacts. (United Nations Human Rights Office of the High Commissioner 2021)

Poor people are more exposed to toxics than rich people and they suffer greater impacts, but everyone who is exposed is at risk.

Those harmed by the toxic releases in Texas and North Carolina will sue when they are able to definitively determine the companies that have caused them harm. And my expectation is that five to ten years from now, some of the financial impact of these lawsuits will be in the news and on our political agenda.

Since many of the impacts of toxic releases are limited to single states, and even single towns and counties, the most realistic approach to reducing the danger from toxics in the United States would be stronger state and local rules. Building codes, zoning, and other land use rules could be employed to provide communities with better protection from releases of toxic substances. Our economic dependence on these businesses will act as a political counterforce to new rules, but it also could open up the possibility of a dialogue among these businesses and their hometowns about what is needed and how to fund it. A state-funded tax credit for new steps taken to reduce toxic releases might be considered. The absence of these policy approaches ensures that toxics from inadequate materials management will continue to degrade the environment.

After a natural disaster, many damaged facilities need to be reconstructed. We saw that in New York after Hurricane Sandy. The point at which damaged facilities are reconstructed is a rare opportunity to rebuild or repair them to meet higher standards of resiliency. In New York, buildings that endured flooded utility rooms moved their boilers and water heaters from basements to higher floors. Hospitals added redundant systems to ensure that they could function if power or water were lost. Tunnels were reconstructed to be capable of withstanding the impact of flooding.

Given the "free enterprise first" culture of North and South Carolina, any improvement in toxic materials management will need the strong support of the business community. While in normal times this is unlikely, the aftermath of a massive flooding event is not a normal time. You hope that you've just experienced a once-in-a-lifetime event, but in the back of your mind, you assume, or at least fear, that extreme weather events are here to stay. This provides an opportunity to encourage, if not require, that more secure and technologically advanced toxic material control practices be put in place.

Pig waste processing might be a good place to start. The construction of facilities to utilize this waste for fertilizer and fuel should be considered. Dumping the waste into ponds is an environmental catastrophe waiting to happen. The argument against effective materials management is that it adds costs to production and makes the business less competitive. But if waste materials are reprocessed for use, and if the production process is designed to reuse most such materials, then waste is transformed from an economic liability to an economic asset. The goal is to make money from your waste; that is the argument for a circular economy. Think of it as chicken wing logic. In the old days, cooks discarded chicken wings, but then a Buffalo bartender served them with hot sauce and gave them away to encourage patrons to buy more beer. The wings grew popular in themselves, going from waste to finger food. But it took imagination and insight to develop that new practice. Mishandling of toxic materials tends to be built on old habits and resistance to change, especially when the motivation for change is a government regulation. But catastrophe can also stimulate new thinking and motivate change.

The mindset in the Northeast after Hurricane Sandy was that reconstruction needed to build on the lessons learned from the hurricane. One of those lessons is that it could happen again. Toxic materials that were released into the environment after a hurricane will be released after the next one too unless we learn how to do a better job of handling them. A natural disaster provides an opportunity to reengineer and reimagine the way that toxic materials are managed. Since we are experiencing greater and greater numbers of high-cost natural disasters, we have a ready source of opportunities to do what President Biden called "build back better."

According to the National Oceanic and Atmospheric Administration (NOAA), the number of extreme weather events is growing:

> In 2021, there were 20 weather/climate disaster events with losses exceeding $1 billion each to affect the United States. These events included 1 drought event, 2 flooding events, 11 severe storm events, 4 tropical cyclone events, 1 wildfire event, and 1 winter storm event. Overall, these events resulted in the deaths of 688 people and had significant economic effects on the areas impacted. The 1980–2021 annual average is 7.4 events (CPI-adjusted); the annual average for the most recent 5 years (2017–2021) is 17.2 events (CPI-adjusted). (NOAA 2021)

While natural disasters can cause problems for manufacturers that are not prepared for them, extreme weather is not the only cause of harm. There are also many routine production processes in both service and manufacturing organizations that do not account for their impact on the environment. Wasted energy and poorly designed waste management systems are perhaps the most common issues in production processes since the costs of energy and waste are simply absorbed into organizational overhead or prices. Wasted materials in manufacturing and discharges of effluent or emissions, while often regulated, are often accepted without question as a necessity of the production process. It is sometimes referred to as "breakage." In many organizations, the philosophy is, "If it ain't broke, don't fix it." If a philosophy of continuous improvement is adopted, waste reduction and more efficient use of energy and materials can be seen as an opportunity to make the organization more cost-effective. As W. Edwards Deming says, "Reliable service reduces costs. Delays and mistakes raise costs" (Cohen and Brand 1990).

Another deficiency in our current production process is that our economic and regulatory models do not require or encourage producer responsibility. As a "to remain anonymous manufacturer" once told me: "Once the product leaves our factory, it is no longer our problem." Nevertheless, the manufacturer of a smartphone is best situated to buy an old phone back in exchange for offering a discount on a new phone. Hewlett-Packard (HP) offers free pickup of used printer cartridges, allowing the company to save money and raw materials by selling refilled cartridges. In New York State, there is a $75 deposit on tires, which reduces the number of tires left on the side of the road and increases the number that are recycled. The producer responsibility model is an example of the circular economy and the opposite of the planned obsolescence model currently imbedded in our linear economy.

DEVELOPMENT OF LAND THAT SHOULD BE PRESERVED

One of the great contributions of President Theodore Roosevelt to America's environment was his active effort to preserve lands largely in the American West. He understood the forces of development, wanting to ensure that forests were used but not overused by humans. Roosevelt was a conservationist, hunter, and early environmentalist. According to the National Park Service:

> The conservation legacy of Theodore Roosevelt is found in the 230 million acres of public lands he helped establish during his presidency. Much of that land—150 million acres—was set aside as national forests. Roosevelt created the present-day USFS in 1905, an organization within the Department of Agriculture. The idea was to conserve forests for continued use. An adamant proponent of utilizing the country's resources, Roosevelt wanted to insure the sustainability of those resources. Roosevelt was also the first president to create a Federal Bird Reserve, and he would establish 51 of these during his administration. These reserves would later become today's national wildlife refuges. (National Park Service n.d.)

In New York State, Governor Nelson Rockefeller ensured that large areas in the Adirondack Mountains were preserved as well, and many states in the United States have established park systems for recreation and preservation. State parks have been established all over the United States. According to Erin E. Williams of the *Washington Post*:

> Across the country, 9,095 state parks span 19 million acres and boast an abundance of natural and cultural riches that can rival those in the

national park system, according to 2019–2020 data from the National Association of State Park Directors. They also highlight local history and are often more accessible. . . . Each state park system is different—some are more developed than others, and infrastructure can include rudimentary campgrounds or concert venues—but together they serve a unique public lands purpose. (Williams 2022)

Despite national, state, and local parks, the pressure on land resources for development for housing, mining, and other types of economic uses is relentless in the United States and is particularly ferocious in large parts of the developing world. Decisions about land use preservation are sometimes as fragile as the lands themselves. There are countless examples of elected leaders pushing development over all other goals. Even in the United States, President Donald Trump decertified national monument lands and permitted the exploitation of federal lands wherever he could. Prioritizing money over nature is nothing new, but it is the underlying cause of the loss of forests and critical ecosystems in the United States and worldwide. The popularity of American park systems demonstrates that we know how to preserve land, and when we do, it can be politically expedient. However, efforts at land preservation often run into political opposition by local and sometimes global economic interests.

THE ABSENCE OF ENVIRONMENTAL VALUES AND ETHICS

At the most fundamental level, the basic cause of environmental degradation is that the value system held by many people does not consider environmental quality to be important. They do not believe that destroying the environment is unethical. We have made tremendous progress cleaning up our environment while growing our economy, but the overly ideological and scientifically illiterate people running many state governments and their counterparts in the Congress are hard at work trying to undermine the successful environmental laws and rules that keep us safe. There are ethical issues at stake that go beyond anything that cost-benefit analyses can tell us. How do you set a price on not poisoning an unborn baby? How do you price the agony of knowing that the neurological damage to a child could have been avoided if the narrow analysis of costs and benefits had been expanded to include a consideration of the ethics of putting profits above people? Even if you are unwilling to conduct a comprehensive analysis of economic costs and benefits, what about the morality of poisoning humans? Why are opponents of renewable energy so eager to burn this filthy fuel when there are cleaner, cheaper alternatives?

In addition to the issue of poisons, there is an environmental ethic rooted in spirituality. Nature was not created by humans, but by a force beyond human capacity and scientific understanding. Under this environmental ethic, it is immoral to destroy nature. The worldview of many people opposing environmental protection these days considers promoting private profit to be the highest purpose and goal of public policy. In this view, nature is simply there to be exploited. Businesses should be allowed to pollute as much as they want, so long as they make money while doing it.

Still, the views of those without a sense of environmental ethics are not universal and may not be a majority. For the young people who will inherit these damaged institutions and our degraded planet, the need for action is urgent and personal. The high school kids protesting for action on climate change understand that they need to shake up the comfortable assumptions of those in power.

As we become more urban, it has also become more important to expose city dwellers to the natural world and build an appreciation of nature. Natural phenomena like the Grand Canyon inspire awe and are considered by some to be evidence of the presence of a deity perhaps beyond human comprehension. A world of disinformation and computer-generated virtual reality could threaten the significance of nature and the ethical imperative of its preservation.

In many respects, the absence of an environmental ethic is probably the most fundamental underlying cause of environmental degradation. Building this sense of connection to the planet and its well-being is a necessary, although not sufficient, condition of effective anti–environmental degradation policies and practices.

POLITICAL PRESSURE FOR RAPID ECONOMIC DEVELOPMENT AND LACK OF UNDERSTANDING OF THE CONNECTION OF ENVIRONMENTAL PROTECTION TO ECONOMIC GROWTH

There are political and business leaders who do not care if economic growth causes environmental damage, and there are environmental advocates who do not believe that you can have economic growth without causing environmental damage. In a *New York Times* piece in 2020 on the climate and economics discussions at Davos, Mark Landler and Somini Sengupta reported:

> Critics pointed to a contradiction that they said the corporate world had been unable to resolve: how to assuage the appetite for economic growth, based on gross domestic product, with the urgent need to check carbon emissions. "It's truly a contradiction," said Johan Rockström, director of the Potsdam Institute for Climate Impact Research. "It's difficult to see if the current G.D.P.-based model of economic growth can go hand-in-hand with rapid cutting of emissions," he said. (Landler and Sengupta 2020)

I find this dialogue a little amazing since it completely ignores the history of America's success in decoupling the growth of gross domestic product (GDP) and the growth of environmental pollution. This fact of American environmental and economic life began around 1980, a decade after the creation of the EPA, and it continues today. It's quite simple: With public policies ranging from command-and-control regulations to direct and indirect government subsidies, businesses and governments developed and applied technologies that reduced pollution while continuing economic growth. This is not a fantasy—it is history. In the 1960s, you could not see the nearby mountains from downtown Los Angeles; today, you can. In the 1960s, you could not ride a bike on a path on the Upper West Side of Manhattan next to the Hudson River; today, you can. Until 1984, New Yorkers dumped raw sewage into the Hudson River; today, with some notable exceptions in rural areas, Americans treat and detoxify our sewage waste. And both Los Angeles and New York City have larger economies in 2022 than they had in 1980. In case you believe that this progress was due to deindustrialization, the two largest sources of air pollution are power plants and motor vehicles, and we have many more of them today than we had in 1980. Both utilize pollution control technology required by regulations and laws.

Environmental protection itself contributes to economic growth. Somebody makes and sells the air pollution control technologies that we put on power plants and motor vehicles. Somebody makes money building all those sewage and water treatment facilities. Just as someone makes a profit selling solar cells and windmills and whoever invents the 1,000-mile-capacity, low-toxic battery that will power electric cars someday will become very, very rich. And environmental amenities are worth money. The cleaner Hudson made the waterfront more suitable for housing development. And the building boom that occurred on New York's West Side followed the cleanup of the Hudson River. An apartment across the street from a park will bring a higher price than the same apartment a block away. The revival of New York's Central Park raised the value of the already high-end real estate on the park's borders. Clean air and water, healthy food, and preserved nature all benefit human health and result in far more economic benefit than economic cost.

The climate crisis is not caused by economic growth, but by the absence of effective public policy designed to reduce greenhouse gas emissions. There is nothing incompatible with capitalism and environmental protection, so long as rules are in place that control the environmental impacts of the products and services that we make and use. With those rules in place, a concern for environmental sustainability can and will permeate everyday decision-making in the private, nonprofit, and governmental organizations from which we all benefit.

While the United States and Europe have decoupled economic growth and pollution, some analysts question the amount of pollution reduction that we have achieved while growing GDP. According to Klaas Lenaerts, Simone Tagliapietra, and Guntram Wolff:

> While globally, CO_2 emissions per unit of GDP are declining, the decoupling rate from 1995 to 2018 was only −1.8 percent annually. To achieve net zero by 2050, the rate would have to accelerate to −8.7 percent, assuming population and GDP growth projections as given, or by a factor

of almost five. To keep GDP growth and population at their projections and thus reject the proposition of de-growth, decoupling would have to accelerate massively. (Lenaerts et al. 2021)

That is, of course, precisely what a decarbonized economy would achieve. The pace of technological displacement and changed consumption patterns will grow, and with it the rate of decoupling.

I've written often about the evolution of the field of management over the past century or so, and stated that a concern for sustainability is the newest trend in the development of more sophisticated organizational management. In the twentieth century, we saw the field of management absorb the development of mass production, social psychology, accounting, and information management. In the twenty-first century, management has dealt with satellite and cellular communications, globalization, and most recently, a concern for the physical dimensions of environmental sustainability. Sustainability managers continue to lead an organization's marketing, strategy, finance, and work processes, but they also seek to assess their use of energy, water, and other materials and work to reduce waste and environmental impacts. Just as finance staff, reinforced by Securities and Exchange Commission (SEC) rules, learned to identify and reduce self-dealing, conflict of interest, and fraud; environmental sustainability staff, reinforced by EPA and SEC environmental risk disclosure rules, look to identify and reduce organizational practices that damage the environment.

On the production side, organizational managers work to increase environmental sustainability, but on the consumption side, consumers are not only buying green, but also changing patterns of consumption that help reduce environmental damage. Going to the gym, riding a bike, and eating a salad are all activities that add to the GDP. But so does taking your private jet to your ski lodge, driving in your internal combustion–powered sport utility vehicle (SUV) to the ski slopes, and eating a steak. All consumption behaviors are not created equal and do not have the same impact on environmental sustainability. More sustainable lifestyles are emerging, and they can be detected in consumption patterns. For example, young Americans seem less interested in owning cars than their older siblings and parents did. Ride-sharing, bike-sharing, and other transit options have become feasible due to the development of the smartphone. But sitting in an Uber and driving your own car are both economic activities that are counted in the GDP (Akhtar 2021).

In the United States, we have seen a dramatic increase in bike sharing. According to Ana Ley of the *New York Times*:

> Since it started in Manhattan and parts of Brooklyn in 2013, Citi Bike, operated by Lyft, has grown from 6,000 bikes and 12,000 parking docks to about 24,000 bikes and 47,000 docks. . . . Bike-share programs are thriving in major cities across the country. In Boston, the program started with about 1,000 bikes and now has 4,000. In San Francisco, a system that started with 700 bikes in 2013 now has 4,500, and during the same period a program in Chicago has grown from 750 bikes to 11,500. (Ley 2021)

These consumption trends are more influenced by changing cultural norms than by public policy, and typically should not be subjects of policymaking. In the case of bike sharing, a public-private partnership is needed to obtain public space for bike docks. Public policy is needed when consumption has a direct negative impact on others, such as driving while intoxicated or smoking in a public space, or when bikes don't observe traffic rules. The environmental impact of consumption can also be reduced by new technologies. For example, streaming music and video have far less environmental impact than videos and discs that need to be manufactured, packaged, and shipped before they are used.

It is ironic that some environmentalists, along with some climate deniers, share the belief that we must trade economic growth for environmental protection. We can and must accomplish both. A reason that we cannot abandon economic development is that most people in the developed world like the way they live and will not give up their way of life. Asking them to do so dooms environmental advocates to political marginalization and failure. In addition, due to the internet, even very poor people in the developing world see the way that many people live in the developed world, want it, and are demanding that their political regimes help them achieve their dreams. The absence of economic development leads to political instability and the potential for violence. Climate scientists often mention the impact of climate change on political instability, and the phenomenon of climate refugees is well documented. But the path to climate mitigation is not through slower economic growth, but through economic growth that is steered toward environmental sustainability and away from gratuitous environmental destruction.

People in the developed world are reluctant to change how they live, and people in the developing world are eager to live like people in the developed world. The political pressure created by the widespread demand for modern technology and economic well-being is fierce and irresistible. The leaders of the world, both democratic and autocratic, feel this pressure and have little choice but to respond to it. This leads me, as always, to the same place: The solution to the problems created by our technology can realistically be solved only by new technology. The heart of the issue is energy and our desperate need to transition as quickly as possible away from fossil fuel use.

One of the first sustainability books I ever read was Ian McHarg's *Design with Nature*. McHarg advocated cluster development as an alternative to suburban sprawl. The idea was that rather than providing every home with a quarter-acre of land and its own large yard, you would build the housing in the one area of the building site that would cause the least damage to natural drainage and ecosystems and preserve the rest of the land as a park for hiking and viewing. It turned out that most of the outdoor access that people used in their homes was on their patios, and suburban yards were not simply ecological disasters—they were a burdensome waste for most homeowners. McHarg demonstrated that with care, humans could build urban developments that might minimize rather than maximize environmental damage.

Sloppy management, the hunger for easy money and short-term profit, and ideological rigidity lead some to believe that the environment must be sacrificed for economic growth. The belief that capitalism is evil and inevitably causes environmental destruction leads others to believe that sustainable economic development is not feasible. My view is that with enlightened design, sustainability management, and cutting-edge technology, we can harness human ingenuity to the practical problems of environmentally sustainable economic development. We can build and live in sustainable cities and end the climate and ecological crises that seem so overwhelming today.

UNDERINVESTMENT IN
ENVIRONMENTAL PROTECTION TECHNOLOGIES

Energy, water supply, and waste management infrastructure tend to be better in rich communities than in poor communities. This is not to say that infrastructure is particularly state-of-the-art anywhere in the United States. For decades, Americans have refused to pay the taxes needed to fund sufficient capital investment in infrastructure. The dire need for reconstruction of our aging infrastructure provides an opportunity to shape the post-COVID-19 economy in a way that improves environmental quality in all neighborhoods, not just those with rich people. Just as FDR brought electricity to rural America, Joe Biden's infrastructure programs could help build a clean and efficient American electrical grid. It could also modernize transportation systems, water supply facilities, sewage treatment plants, and systems of solid waste management. The American infrastructure revitalization program enacted in late 2021 put people to work while improving the efficiency and quality of our communities and economy.

The good news about most infrastructure investment is that it tends to have a multiplier effect by making the economy more efficient. A rehabilitated energy or water system can reduce wasted energy and water. A new port can save the time and money it takes to ship goods. Better mass transit can reduce the cost of moving people from place to place. Better roads and bridges allow motor vehicles to transport people and goods more efficiently.

CONSUMER DEMAND FOR PRODUCTS THAT POLLUTE
AND OUR SEDUCTIVE LIFESTYLES

An underlying cause of environmental degradation is that many people like their lifestyles and continue to advance consumer demands for products that pollute. From plastic drinking straws to SUVS, people like this stuff, and as long as they maintain these preferences, corporations will work to meet their demands. Asking people to reduce their consumption is a futile political tactic that typically engenders resistance. That does not mean that we shouldn't educate people about products that create harm. That has been done now for over half a century for tobacco, and still over 19 percent of the world's adults smoke, and tobacco use causes over 7,000,000 deaths a year. Plastic and toxic remains from discarded cigarette butts is a major source of urban litter and pollution. The percentage of people who smoke is going down, but it is impossible to ban smoking.

Products that cause environmental harm should be identified, along with products that create less harm, but the demand for harmful products will not go away. Not only will people who consume these products insist on continuing to do so, but due to the low cost of communication, images of this consumption are transmitted globally and create wants and demands throughout the world. Advertising, product placement in movies, and images of trendsetters on social media create demand for the image (and reality) of wealth and the glitzy lifestyles that some rich people pursue. These images, central to planned obsolescence and our current linear economy, are growing in number and present a great challenge to environmental sustainability.

IGNORANCE OF SCIENCE AND ENVIRONMENTAL IMPACTS

Today's decision-makers, having never experienced the global catastrophes and world wars of the twentieth century, seem unable to think beyond the next election cycle and the next quarterly report. The horrific invasion of Ukraine by Russia in early 2022 may have reawakened awareness of this potential for catastrophe. Russia's attack on Ukraine's nuclear power plants reopened discussions about the dangers of radioactivity. The debate about the impact of technologically induced environmental impacts is central to American democracy and is badly served by ideologically tainted media, ranging from Fox News to MSNBC, and from the *Drudge Report* to the *Huffington Post*. The media has become an echo chamber that reinforces biases and hardens ideologies.

The impact of our lack of science literacy can be seen in decision-making in the Trump White House and in regulatory agencies like the

EPA that aggressively resisted science when he was president. We saw that at all levels of government during the COVID crisis of 2020–2022. Instead of debating the validity of health findings on scientific grounds, some people rejected environmental and health science entirely. This exacerbates our science literacy problem when scientific findings are considered something to believe in rather than something to learn and understand. This is not to argue that science is always fact based and value neutral. Science is not without value choices and ideologies. The problems that scientists choose to study reflect what they consider important and what they consider to be important reflects their values. An earth scientist's values may simply be the advance of human knowledge about how the Earth works and may have little connection to any concerns about the impact of humans on the planet's well-being. But even the goal of advancing knowledge must also be considered a value choice. While values play a role in science, the scientific method places a high value on the role of measurement and observations. In other words, high value is placed on verified facts and observations. Good science tries to reduce but can never eliminate bias and value-laden perspectives.

But the impact of science and engineering on our lifestyle cannot be overstated. Civilization once replaced human labor with animal labor, and then replaced animals with mechanical labor, and now we are in a world of computer-controlled electronic automation. Everything around us, from our refrigerators to our autos, is controlled by technologies that we use but do not understand. We live in a global and interdependent world, reliant on technology and scientific experts for our survival. Corporate decision-making and public policymaking require a growing degree of input from scientific and technical experts. While the expertise comes from engineers and physical and biological sciences, the decisions are largely made by people trained in law or business. Although the lawyers are sometimes experts in some element of law and the businesspeople often have some understanding of finance and the underlying math, most senior leaders are a long way from their days of focused, analytic inquiry. They depend on communicating with experts but do not always know how to do so.

At the start of my professional career when I worked at the EPA, I was struck by the relative lack of scientific expertise among decision-makers. I had been hired in part to staff an organizational analysis and develop policy guidance for citizen participation in Superfund toxic waste clean-ups. I had focused my doctoral studies on political science and public policy on organization theory, and I wrote my dissertation on public participation in environmental decision-making. Much of my work at the EPA was within my areas of expertise. But then, at times, I was asked to participate in policy analyses that required an understanding of toxicology, hydrology, soil science, and engineering. Needless to say, I learned the difference between something I was an expert in and something I was not. That experience was humbling and taught me to stay in my lane and focus my work on something that I truly understood. It also taught me the importance of learning to work across disciplines and in teams. We could not address environmental problems without drawing on many forms of expertise. Environmental policy and management require people who understand organizations, law, finance, and business, but we also need engineers, environmental scientists, experts in public health, and many other types of specialists.

But decision-makers do not have the luxury that consultants, policy analysts, and academics enjoy. They often don't have time to search for expertise and slowly learn from experts. They need experts and expertise, but they often need them and their knowledge in a hurry. Sometimes one must fly the jet plane and repair it at the same time. A recession looms, and lawyers must learn finance. A pandemic hits, and all of us must learn about the science of virus transmission. It is important that we have enough scientific background to know what we don't know. For those who are experts in one area, it is critical to recognize that they need to consult with someone who is an expert in another. For those who are not experts in a subject matter, but professionals trained in the methods of management, quantitative analysis, communication, financial analysis, law, or some other field, it is essential that they learn the critical importance of scientific expertise. They must not only know what they don't know, but also learn how to learn what they need to know to make decisions. They need to learn how to elicit the views of competing experts and identify areas of scientific consensus that mandate or justify action. This requires setting aside personal biases and, during a crisis, personal interest.

There is little question that the White House under President Trump, as well as many state and local governments during the height of the COVID crisis, conducted a master class in the cost of scientific illiteracy. At the start of the COVID crisis, medical and public health expertise was heard for a very short time in the White House briefing room, but it was quickly overwhelmed by Trump's political rhetoric. Eventually, health experts were exiled after the president proposed the Clorox cure in one press conference. A president with Trump's degree of scientific illiteracy and ego could simply not share a platform with world-class medical experts.

Unlike many other problems faced by political leaders, the facts of COVID-19 could not be denied. Politicos in and out of the White House worked hard to spin and minimize the danger, politicize masks, and label prudent public health measures as attacks on liberty. One major success of the Trump administration was "Operation Warp Speed," which resulted in the development of effective vaccines in record time. Unfortunately, once vaccines were available, the political debate shifted to vaccine mandates versus personal freedom. The impact of this willful scientific illiteracy was massive illness and death—along with economic turmoil. As we came out of the pandemic, health experts were on the defensive and gave confusing and conflicting advice. This further undermined the role of science and scientists in policymaking.

We have seen similar problems with the science of climate change (Funk and Kennedy 2020). The facts and impacts of climate change are irrefutable at this point. What we do about it and how we address it are subject to debate about policies and management approaches. Climate scientists sometimes seem to believe that the certainty of science should lead to equally certain policy solutions, but public policy is much more complicated than physics. The science of biodiversity is complex, but it can be translated for decision-makers. But it too may lead to policy trade-off decisions. There are many other areas of science that touch on key issues of public and private management. Decision-makers must understand both the science of the problem and the science of the proposed solutions.

Scientists must learn that policy formulation and implementation amount to a craft, not a science.

Scientific facts sometimes must be balanced against other scientific facts, as well as nonscience-based costs and benefits. During the COVID pandemic, the damage of closed schools to child development and to the return to normal economic life was clearly articulated by experts in the physical and mental health of children and by economists. On the other side of the debate, we saw medical and public health experts seeking to understand this new disease and reduce its impact on human health. The stakes were high, and it is very difficult to balance the claims of experts from these diverse disciplines. Moreover, decisions on school openings are not only made by government, but also by parents who may decide to keep their children at home.

The COVID pandemic was an example of months of public decisions that needed to be made in the face of scientific uncertainty, by decision-makers who did not know what they didn't know and didn't fully understand the range of options open to them. The opportunities to contain the virus in China, Europe, and the United States were missed by governmental decision-makers in each place. The methods used to reduce virus transmission have worked, but the economic cost of those methods resulted in reopenings that caused the virus to come back to places that had driven it out.

To maintain "normal" life and avoid the next pandemic or the next extreme weather event, we will need to do a better job of utilizing scientific expertise to address the negative impacts of our technologies. Climate denial and pandemic denial are examples of willful scientific ignorance. In a world as dependent on technology as the one we live in, we should expect that technologies from air travel to food production will carry dangers that we must learn to identify and mitigate. We listen to experts when they invent something new that we think we will like, but we ignore them when they deliver bad news about negative impacts. We need a more mature and sophisticated approach to utilizing scientific expertise in decision-making. COVID-19 highlighted the seriousness of the impact of our choices: Illness, death, and economic destruction were widespread and could have been much less damaging. There are many less dramatic examples of the impact of scientific illiteracy on our society and economy. In this case, ignorance was far from bliss.

CONCLUSION

This chapter focused on the fundamental or underlying causes of environmental degradation. We have developed institutions, patterns of land use, economic systems, and political processes that are ill equipped to deal with the impact of the technological complexity that we benefit from and rely on. The benefits that we enjoy and the way of life that we benefit from do not come without costs. When these costs are visible and the harm is immediate, we are capable of responding. But we'd all prefer to enjoy the benefits without paying the costs. The crisis of environmental sustainability has become visible and requires a response. But that response must embody a workable strategy that expands rather than contracts support for a cleaner, more sustainable environment. Chapter 4 presents such a strategy and is in many respects, the heart of my argument.

WORKS CITED

Akhtar, Rulia, Sayema Sultana, Muhammad Mehedi Masud, Nusrat Jafrin, and Abdullah Al-Mamun. 2021. "Consumers' Environmental Ethics, Willingness, and Green Consumerism between Lower and Higher Income Groups." *Resources, Conservation and Recycling* 168 (May): 105274. https://doi.org/10.1016/j.resconrec.2020.105274.

Anthes, Emily. 2022. "A Silent Victim: How Nature Becomes a Casualty of War." *New York Times*. April 13, 2022. https://www.nytimes.com/2022/04/13/science/war-environmental-impact-ukraine.html.

Brumfiel, Geoff, Meredith Rizzo, Tien Le, and Alyson Hurt. 2022. "Video Analysis Reveals Russian Attack on Ukrainian Nuclear Plant Veered near Disaster." NPR, March 11, 2022. https://www.npr.org/2022/03/11/1085427380/ukraine-nuclear-power-plant-zaporizhzhia.

Cohen, Steven, and Ronald Brand. 1990. "Total Quality Management in the U.S. Environmental Protection Agency." *Public Productivity & Management Review* 14 (1): 99–114. https://doi.org/10.2307/3380525.

Funk, Carly, and Brian Kennedy. 2020. "How Americans See Climate Change and the Environment in 7 Charts." Pew Research Center, April 21, 2020. https://www.pewresearch.org/fact-tank/2020/04/21/how-americans-see-climate-change-and-the-environment-in-7-charts/.

Government Accountability Office (GAO). 2019. "Superfund: EPA Should Take Additional Action to Manage Risks from Climate Change," October 18, 2019. https://www.gao.gov/products/gao-20-73.

Krauss, Clifford, Manny Fernandez, Ivan Penn, and Rick Rojas. 2021. "How Texas' Drive for Energy Independence Set It Up for Disaster." *New York Times*, February 21, 2021. https://www.nytimes.com/2021/02/21/us/texas-electricity-ercot-blackouts.html.

Landler, Mark, and Somini Sengupta. 2020. "Trump and the Teenager: A Climate Showdown at Davos." *New York Times*, January 21, 2020. https://www.nytimes.com/2020/01/21/climate/greta-thunberg-trump-davos.html.

Lenaerts, Klaas, Simone Tagliapietra, and Guntram Wolff. 2021. "Can Climate Change Be Tackled without Ditching Economic Growth?" *Bruegel*, September 16, 2021. https://www.bruegel.org/2021/09/can-climate-change-be-tackled-without-ditching-economic-growth/.

Ley, Ana. 2021. "Citi Bike Struggles to Keep Up with New Yorkers' Love of Cycling." *New York Times*, December 2, 2021. https://www.nytimes.com/2021/12/02/nyregion/citi-bike-parking-docking-station.html?searchResultPosition=1.

National Oceanic and Atmospheric Administration (NOAA). 2022. "Billion-Dollar Weather and Climate Disasters." https://www.ncdc.noaa.gov/billions/.

National Park Service. n.d. "Theodore Roosevelt and Conservation." Accessed March 31, 2022, from https://www.nps.gov/thro/learn/historyculture/theodore-roosevelt-and-conservation.htm.

Reals, Tucker. 2022. "Ukraine Says Russia's Putin Has 'Ordered the Preparation of a Terrorist Attack' on Chernobyl Nuclear Plant." *CBS News*, March 11, 2022. https://www.cbsnews.com/news/ukraine-chernobyl-russia-putin-orders-terrorist-attack-nuclear-plant-kyiv-says/.

Tabuchi, Hiroko. 2021. "Lack of Power Hinders Assessment of Toxic Pollution Caused by Ida." *New York Times*, September 1, 2021. https://www.nytimes.com/2021/09/01/climate/hurricane-ida-toxic-pollution.html.

Thrush, Glenn, and Kendra Pierre-Louis. 2018. "Florence's Floodwaters Breach Defenses at Duke Energy Plant, Sending Toxic Coal Ash Into River." *New York Times*, https://www.nytimes.com/2018/09/21/climate/florences-floodwaters-breach-defenses-at-power-plant-prompting-shutdown.html.

United Nations Human Rights Office of the High Commissioner. 2021. "Vulnerable Groups and Toxic Exposures." https://www.ohchr.org/en/special-procedures/sr-toxics-and-human-rights/vulnerable-groups-and-toxic-exposures.

Williams, Erin E. 2022. "A Pivotal Period: Century-Old State Park Systems Face Modern Issues." *Washington Post*, April 7, 2022. https://www.washingtonpost.com/travel/2022/04/07/state-park-systems-future-challenges/.

World Nuclear Association. 2022. "Nuclear Power in Ukraine," June 2022. https://world-nuclear.org/information-library/country-profiles/countries-t-z/ukraine.aspx.

4

A STRATEGY FOR REDUCING POLLUTION AND GROWING A RENEWABLE RESOURCE-BASED ECONOMY

Earlier chapters of this book have defined environmental sustainability and summarized the current state of environmental degradation and the causes of our degrading environment. With an overview of our environmental conditions in hand and a rudimentary understanding of how and why they've been created, we are ready to develop an approach to achieving environmental sustainability. This chapter presents a realistic, pragmatic strategy for reducing pollution and growing a renewable resource–based economy. A key element of the strategy is providing public incentives for clean production and consumption. To do that, we need to clearly define a clean economy by developing generally accepted sustainability metrics. We then need to measure current conditions rigorously and work steadily to improve them.

The key is to connect environmental sustainability to economic growth. As I've mentioned throughout this book so far, we need to discard the concept that environmental protection can take place only through reduced economic consumption. Production and consumption must change, and elements of a circular economy need to be implemented. Government will need to play a role in this transition by structuring regulation and taxation to encourage and reward renewable resource use, recycling, sharing, and pollution prevention. Central to this strategy is a positive focus on promoting renewable energy rather than a negative focus on opposing fossil fuels. The goal is to accelerate the process of driving fossil fuels from the competitive marketplace. But even with government intervention, this will be a generation-long transition.

At the local level, we need to implement elements of the circular economy. All of this will also require the use of procurement policy and infrastructure investment by the federal and eventually state and local governments to drive capital toward the green economy. In addition, government authority must be used to ensure that the exploitation of undeveloped land is minimized and, where undertaken, adheres to the principles of environmental sustainability.

The general approach draws on parts of the Green New Deal. The Green New Deal was an effort by progressive politicians to link income redistribution, economic development, and environmental protection. A Green New Deal could mirror FDR's series of policy innovation experiments during the Great Depression. While ambitious targets are useful, decarbonizing the American economy by 2030, as some Green New Deal plans promote, is infeasible and would disrupt our way of life. My view is that one key infrastructure need is in modernizing the electrical grid and building microgrids that can be knitted together into a smart grid but kept distinct to enhance resiliency. Transportation infrastructure should include electric motor vehicle charging stations, which generally should be private but stimulated with public funding. It should also include mass transit systems, including car and ride sharing, electric buses, light rail, and rail. We will also need to invest in the infrastructure needed to protect us from the impact of the climate change that we have already baked into the atmosphere. Our shorelines need protection. The flood controls that we set up to protect our farms and river towns must be strengthened and rebuilt. The key research needs are in energy storage and solar cells. The cells and batteries must become lighter, smaller, cheaper, and more efficient. They also need to be less toxic and less dependent on rare earth materials. Further, we need to develop automated methods of waste sorting to enable us to mine resources from our waste stream.

The overall strategy is to use incentives and fund research that will develop sustainability technologies and practices and help put them in place. Over the past century and a half, the American economy has steadily been transformed by the development and diffusion of new technologies: steam engines, the electric grid, light bulbs, heating, ventilation, and air conditioning (HVAC) systems, refrigeration, air conditioning, radio, television, the internet, global positioning systems (GPS), smartphones, the internal combustion engine—the list is long. The pace of innovation has largely been determined by the free market. Despite market dominance, the federal government has often intervened, largely through the defense budget, and in the process created a world of low-cost communication and information. It also intervened to promote home ownership with guaranteed mortgages and income tax deductions for property taxes and mortgage interest, and to promote automobile use by building roads and highways. Environmental sustainability needs the same creative, energetic use of governmental authority and resources. We need to self-consciously accelerate the development and use of technologies that enable economic growth while preventing environmental damage.

PROVIDING PUBLIC INCENTIVES AND POLICIES PROMOTING CLEAN PRODUCTION AND CONSUMPTION

President Joe Biden established his trillion-dollar infrastructure package to help rebuild and enhance the nation's transportation, energy, and water systems. For this to work, the federal government's investment needs to be more than matched over the next decade by private investments and investments by state and local governments. Even more federal funding will also be required. The transition to environmental sustainability will be built on public-private partnerships that invest in environmentally sustainable economic growth. The complexity of the technology, economic relations, and environmental impacts of the global economy requires an active, responsive, and competent federal government. The modern economy is mixed, with both public and private players. The sooner ideologues get past glorifying one sector over the other, the sooner we can begin to develop the partnerships that are really required. Some parts of our economy require public subsidies; others do not. Some policy objectives, like affordable housing and ending childhood poverty, are not going to attract much private investment. But others, like a low-cost, reliable system of renewable energy generation and distribution, could very well attract a boatload of private money, and projects in this area should be designed to do that.

Our goal should be to transform fossil fuel companies to energy companies where management takes seriously the idea that a company's future requires that it redefine itself. Such newly redefined energy companies should apply their engineering, procurement, logistics, and management talents to renewable energy. They should begin a transition to a new set of energy products. They might look to acquire some companies in the renewable energy business and develop a long-term plan for disinvesting in and even closing their fossil fuel extraction and refining businesses.

Fossil fuel companies have reaped the benefits from government policies such as the interstate highway system and the oil depletion tax allowance. Increased mobility was promoted by public policy, and both the public and oil companies have benefited from those policies. But climate change has altered the political environment. The days of fossil fuel subsidies are ending, and they are being replaced by subsidies for their competitors: renewable energy. Decarbonization targets by the government are a direct attack on industries that sell fuels emitting greenhouse gases.

There are a wide variety of incentives, regulations, disincentives, and policies that government can establish to promote a clean and circular economy (European Commission, Directorate General for Research and Innovation 2021). Congestion pricing can be used to tax auto use while subsidizing mass transit. Companies can be given a tax deduction for buying back consumer products. The companies or the government can impose a deposit on computers, television sets, smartphones, and any other consumer product that is repaid to consumers when they return these products to the company. That will encourage the design of products that can be easily disassembled and mined for raw materials. Companies should be rewarded by developing business models built on reusing products. Effluents and emissions should continue to be subject to command and control regulations, with the ability to buy and sell pollution allowances.

Consumer behavior can also be influenced with public policies that incentivize certain types of consumption. Electric vehicles can be encouraged by tax credits. Mass transit can be encouraged by building it and subsidizing its cost for riders. Charging shoppers for plastic bags and offering deposits on the return of plastic bottles can encourage substitutes and recycling. Nonbiodegradable materials should be banned from use in cigarette filters. The installation of solar cells can be subsidized directly or capitalized in return for a share of energy cost savings.

GOVERNMENT PURCHASING, LAND MANAGEMENT, AND INFRASTRUCTURE INVESTMENT

Another element of the strategy for moving to a green economy is to invest federal infrastructure dollars to support the transition and to utilize the federal government's vast purchasing power to help create a market for electric vehicles and renewable energy. We saw both of these elements put into play by the Biden administration in 2021. I will discuss in detail the specific areas of infrastructure investment needed in chapter 5; here, I am addressing the overall benefit of federal spending in the green economy. In addition to its purchasing power, the federal government controls over a quarter of all the land in the United States. Conservation goals can be achieved via enlightened and careful management of that land.

On November 15, 2021, President Biden signed into law a trillion-dollar infrastructure bill that included about $300 million in funding for environmental projects. According to the White House fact sheet on the bill, it included the following environmental provisions:

> The bill invests $105 billion in new funding to improve healthy, sustainable transportation options for millions of Americans by modernizing and expanding transit and rail networks across the country. It will replace thousands of transit vehicles, including buses, with clean, zero emission vehicles. . . . The bill invests $7.5 billion to build out the first-ever national network of EV chargers in the United States. . . . The deal invests $5 billion in zero and low emission school buses, in addition to more than $5 billion in funding for public transit agencies to adopt low- and no-emissions buses. . . . The deal makes our communities safer and our infrastructure more resilient to the impacts of climate change and cyber attacks, with an investment of roughly

$50 billion. This includes funds to protect against droughts and floods, in addition to a major investment in weatherization. The bill is the largest investment in the resilience of physical and natural systems in American history. . . . The deal's $55 billion investment represents the largest investment in clean drinking water in American history, including dedicated funding to replace lead service lines and the dangerous chemical PFAS (per- and polyfluoroalkyl). It will replace all the nation's lead pipes and service lines. . . . The deal invests $21 billion in environmental remediation, making the largest investment in addressing the legacy pollution that harms the public health of communities and neighborhoods in American history. . . . The bill includes funds to clean up superfund and brownfield sites, reclaim abandoned mine land and cap orphaned gas wells. . . . The deal's roughly $60 billion investment is the single largest investment in clean energy transmission in American history. It upgrades our power infrastructure, including by building thousands of miles of new, resilient transmission lines to facilitate the expansion of renewable energy. It creates a new Grid Deployment Authority, invests in research and development for advanced transmission and electricity distribution technologies, and promotes smart grid technologies that deliver flexibility and resilience. (White House 2021a)

In case you don't have your calculator handy, that is $308.5 billion spent on accelerating the transition to a green economy, adapting to climate change, protecting our water supply, and cleaning up the environment. That is a lot of cash that is having a transformative impact on America's economy and environment.

Following that bill, in early December 2021, President Biden took major executive action to move the power of federal procurement behind the green economy. According to a White House fact sheet:

The President's executive order directs the federal government to use its scale and procurement power to achieve five ambitious goals:

- 100 percent carbon pollution-free electricity by 2030, at least half of which will be locally supplied clean energy to meet 24/7 demand;

- 100 percent zero-emission vehicle acquisitions by 2035, including 100 percent zero-emission light-duty vehicle acquisitions by 2027;
- Net-zero emissions from federal procurement no later than 2050, including a Buy Clean policy to promote use of construction materials with lower embodied emissions;
- A net-zero emissions building portfolio by 2045, including a 50 percent emissions reduction by 2032; and
- Net-zero emissions from overall federal operations by 2050, including a 65 percent emissions reduction by 2030. (White House 2021b)

The federal government's purchasing power is enormous, and an executive order like this has a substantial impact on businesses seeking to make and sell green products. As the fact sheet notes, "By transforming how the federal government builds, buys, and manages its assets and operations, the federal government will support the growth of America's clean energy and clean technology industries" (White House 2021b).

The federal government owns 300,000 buildings and 600,000 vehicles. Its nonpersonnel budget is about $650 billion a year. The buildings all require energy to operate, and the vehicles are routinely replaced. While it will take decades to decarbonize the federal government, its enormous force in the marketplace will accelerate the profitability of companies in the renewable energy, energy efficiency, and electric vehicle business. In addition, the federal government controls a huge amount of land that can be managed according to the principles of environmental sustainability. According to the Congressional Research Service:

The federal government owns roughly 640 million acres, about 28 percent of the 2.27 billion acres of land in the United States. Four major federal land management agencies administer 606.5 million acres of this land (as of September 30, 2018). They are the Bureau of Land Management (BLM), Fish and Wildlife Service (FWS), and National Park Service (NPS) in the Department of the Interior (DOI) and the Forest Service (FS) in the Department of Agriculture. A fifth agency, the Department of Defense (excluding the U.S. Army Corps of Engineers), administers 8.8 million acres in the United States (as of September 30, 2017), consisting of military bases, training ranges, and more. Together, the five agencies manage about

615.3 million acres, or 27 percent of the U.S. land base. Many other agencies administer the remaining federal acreage. The lands administered by the four major agencies are managed for many purposes, primarily related to preservation, recreation, and development of natural resources. (Vincent et al. 2021)

While many of the world's worst land management practices take place in the developing world, the United States has had more than its fair share of land use development ideas that have not been terribly well thought out. A key part of our environmental sustainability strategy must be to carefully steward any land under government control.

Although Biden's green procurement executive order exempts operations directly related to national security and intelligence gathering, the Department of Defense has been working on climate adaptation since 2014, and for the past decade, environmental sustainability has been gradually integrated into military operations and planning (Cho 2017). In the summer and fall of 2020, I advised a group of graduate students at Columbia University who were engaged in a management simulation of a proposal to decarbonize the noncombat elements of the Department of Defense (Patten et al. 2021).

My students demonstrated the feasibility of this initiative. In fact, the military pioneered the use of solar energy during the war in Iraq as a way of reducing the use of oil tank trucks, which were highly vulnerable to roadside improvised explosive devices.

In addition to carbon reduction, the executive order required agencies to increase water efficiency, reduce and recycle waste, prevent pollution, and focus on the environmental sustainability of its supply chains. It also required each agency to submit annual plans reporting environmental progress, and perhaps most significantly, it mandates the creation of organizational capacity in the federal government to manage the implementation of President Biden's executive order. Section 501 of the order says:

The Office of the Federal Chief Sustainability Officer is reestablished within CEQ. The EPA shall provide funding and administrative support for the Office.

(a) The Office shall be headed by a Federal Chief Sustainability Officer, who shall be appointed by the President. The Federal Chief Sustainability Officer shall lead the development of policies, programs, and partnerships to achieve the policies set forth in this order, advance sustainability and climate resilient Federal operations, and ensure the Federal Government leads by example in combating the climate crisis.

(b) The heads of all agencies shall cooperate with the Federal Chief Sustainability Officer and provide such information, support, and assistance as the Federal Chief Sustainability Officer may request, as appropriate and consistent with applicable law. (White House 2021b)

Section 502 of the order required each federal agency to appoint an agency chief sustainability officer, and the order included other requirements for the Office of Management and Budget and other parts of the federal government to ensure that its stipulations would be implemented.

In my view, this executive order mandated sustainability management in the federal government. Just as agencies must submit and manage their financial resources through a highly institutionalized budget planning and management process, they will now be expected to manage their physical resources according to principles of environmental sustainability. We will know that this is real if the agency chief sustainability officers come to be seen as senior managers within their agencies. The executive order and its implementation design were excellent first steps, but they will be meaningful only if the president and his inner circle of advisors take them seriously and hold agencies accountable for their performance.

In sum, 2021 represented a turning point in the transition to a green economy in the United States. The infrastructure bill and federal procurement policy began the process of driving many billions of dollars toward renewable energy, energy efficiency, grid modernization, and water and transportation infrastructure. The federal government's impact on the American economy is substantial, and green businesses that might have struggled to start up could be certain to have one very large customer.

There is a tendency in some organizations to see environmental sustainability as less than serious—to view it as public relations rather than an element of operations management. Despite these sustainability imposters, many organizations have come to recognize that inefficient or expensive energy, unmanaged waste, and environmental liability can be major drags on organizational performance and corporate profitability. In some organizations, chief sustainability officers are not seen as important. In others, however, they are central players in organizational strategy formulation and decision making.

Despite its slow-moving, process-focused behavior, the federal government tends to take rules and executive orders seriously. There have been visible efforts to comply with Biden's Green Executive Order. But we will need to look at output and outcome performance measures in the coming years to see whether anything is really going on. It should be easy to count the number of electric vehicles purchased. Energy audits can be performed on government buildings, and it will be possible to track changes in procurement practices. Waste and recycling data is also relatively easy to record and analyze. Data on land use practices and environmental impacts on public lands can also be collected and analyzed.

It will be more difficult to evaluate the impact of Biden's order on the rest of the economy. The 2021 trillion-dollar infrastructure law, combined with federal green procurement, has resulted in a huge infusion of cash into the green economy. We won't know how much this sector would have grown without federal spending, and it will be difficult to disentangle the various sources of federal funding. But I suspect that there will be synergistic effects as the force of federal spending is interpreted by private-sector managers and investors as they make decisions on corporate expenditures. The rapid commitment of major capital investment in electric vehicles by American automakers is an example of the potential impact of Biden's green agenda.

We have begun to see Biden's executive order imitated by state and local governments. Even before the federal actions, we were seeing corporations and large institutions announce carbon reduction goals. The growth of corporate sustainability reporting and finance suggests that Biden's order might well survive a conservative Republican administration.

Capital markets have finally figured out that corporations are not immune to environmental risk. Climate-induced drought and extreme weather can disrupt operations. Toxic substances and invasive species can harm ecosystems, and when contagious viruses are involved, they can bring economies screeching to a halt as they disrupt supply chains, which turn out to be less durable than we had thought. Some wealthy people and public pension funds are insisting upon green investments. We are also learning that just as companies need to pay attention to financial and reputational risk, they must also understand and manage their environmental risks. In 2022, the Securities and Exchange Commission (SEC) published a detailed proposed climate disclosure rule—an acknowledgment of the growing importance of climate risk.

These governmental and corporate investments are a central element of the strategy to transition to a green economy. Government has steered its capital budget toward environmental infrastructure, and its purchasing power has been steered in the same direction.

DEVELOPING SUSTAINABILITY METRICS

In the case of environmental sustainability, I believe that we need to develop a set of generally accepted environmental sustainability indicators. Measures of energy use, water use, material use, waste management/product reuse, pollution, and greenhouse gas emissions should be developed for service and manufacturing organizations. Just as financial indicators have been refined by government over time, so should these indicators. Sustainability units in corporations could be charged with reporting on these measures, and the nongovernmental organizations (NGOs) and private firms now producing corporate sustainability reports could be trained and certified by the federal government to audit these reports. Standard indicators would be developed and reported.

Standardized sustainability metrics are useful not only for creating more efficient regulation, but also for guiding investor decision-making on climate risk. According to a recent press release by the SEC:

The Securities and Exchange Commission today proposed rule changes that would require registrants to include certain climate-related disclosures in their registration statements and periodic reports, including information about climate-related risks that are reasonably likely to have a material impact on their business, results of operations, or financial condition, and certain climate-related financial statement metrics in a note to their audited financial statements. The required information about climate-related risks also would include disclosure of a registrant's greenhouse gas emissions, which have become a commonly used metric to assess a registrant's exposure to such risks. (Securities and Exchange Commission 2022)

I am delighted to see this recognition of the material impact of one of the many financial risks posed by environmental conditions. When Wall Street pays attention, we know that serious money is at stake. There will certainly be ideologues in the U.S. Congress and editorial writers in the *Wall Street Journal* who believe that these risks are overstated and simply being pushed by an opposing ideology. It's true that there is an ideological component to the climate finance debate. Still, investors must be provided with information to assess and understand the exposure to risk in a company's external setting. Some of these risks relate to market conditions, and some relate to social, cultural, political, and environmental conditions. Companies that abandoned their Russian operations in the face of Vladimir Putin's cruel and senseless invasion of Ukraine had to disclose the losses they sustained and the prospects of recovery in their financial reports. Politics creates financial risk, and our ecological environment creates financial risk as well.

My only objection to the SEC proposal is that it is limited to climate risk and does not encompass the full range of environmental risk. The need for a broader framework of environmental sustainability metrics was highlighted in a 2022 *Wall Street Journal* interview conducted by reporter Ed Ballard with Alison Bewick, the head of risk management at Nestlé. According to Ballard:

A Nestlé SA executive who helped put together a new framework for biodiversity reporting said that companies should release integrated disclosures related to climate change and nature, because the two things are so interconnected. Alison Bewick, head of risk management at Nestlé, was one of the executives involved in creating the initial framework from the Taskforce on Nature-related Financial Disclosures that was published last week. The framework, devised by businesses working in collaboration with scientific organizations and nonprofit sustainability standard-setters, is meant to serve as a guide for companies about reporting on nature-related risks and opportunities. It follows the model of the climate-risk framework devised by the Task Force on Climate-Related Financial Disclosures. (Ballard 2022)

As Bewick clearly understands, the overall issue is environmental risk. Climate change is seen by some as the most important risk and an existential threat as well, but it's a little silly to hold a contest between environmental risks. At any one time, any number of risks could threaten us. In 2022, many of us started to think about the risk of radioactive contamination from nuclear power plants damaged by war. At that time, we also continued to live through the risk posed by the invasive COVID-19 virus. There are no shortages of environmental risks caused by the unanticipated impacts of modern technology. Nestlé's Bewick concretely calls for integrating the biodiversity measurement and disclosure framework with the climate framework. In the *Wall Street Journal* interview, she observed:

> When we think about how we can address our carbon footprint, a lot of it's through nature-based solutions. It's beyond just the greenhouse-gas measurement, it's around the availability of water, it could be the soil profile, how you approach land-use in terms of rotation of crops, that type of thing. I think the underlying principle is that this should be ultimately an integrated disclosure, because there's a very strong interconnectivity and dependency between nature and climate. (Ballard 2022)

The resistance to climate science that we see in the political world and in fossil fuel companies reminds me of the resistance to the connection that medical science was making between smoking tobacco and cancer (especially lung cancer, but others as well). The relationship is clear and has been established for many years, but economic interests continue to dominate health concerns. In 2019, 1.1 billion people smoked, and 7.7 million people died from tobacco use. So much for science. . . . Climate change is similar, and if anything, the economic interests threatened are wider and far more powerful than the tobacco industry. Perhaps that is why climate change is such a dominant environmental issue. Mitigating climate change requires fundamental changes in the technologies that drive our economic system.

Climate science is relatively straightforward, and some of the impacts of climate change are well understood. But at a certain point, the relatively simple physics of climate change intersect with far more complex biological and ecological systems. Those changes and the damage to ecosystems caused by nonclimate-related human impacts are not as well understood and are far more difficult to measure. The web of relationships in the living world of ecology is more subtle and complex than the massive impact that greenhouse gases have on our climate. And yet millions of subtle changes to our biosphere can add up to an impact that is easily as massive as that caused by climate change.

Bewick's call for integrating climate and biodiversity measures in a single framework makes sense because the two sets of impacts are interconnected. It is also a way for the relatively less popular biodiversity impacts to cash in on the currency and notoriety of climate impacts. What is most important is that we get beyond this improvisational stage in environmental sustainability metrics. In the world of corporate finance, accounting terms are defined and regulated by government, not by NGOs. When the SEC began during the New Deal, it was responsible for the development of Generally Accepted Accounting Practices (GAAP). The SEC or some other part of the federal government needs to start the process of developing Generally Accepted Environmental Sustainability Metrics. If these are to become routine elements of corporate disclosure, companies need clear definitions of what they must disclose. This might begin with the climate disclosures now proposed, but it then should expand into broader measures of environmental impact and risk.

The aim of the SEC rule is precisely to provide clearer metrics for climate disclosure. According to Richard Vanderford's report in the *Wall Street Journal*:

> The [SEC Climate Disclosure] rule is meant to bring order to what has been uneven climate reporting by different public companies. In place of voluntary sustainability reports which use handpicked metrics, companies would have to disclose in much greater detail how much carbon they emit and how they plan to address looming climate risks. In theory,

investors could then make more informed comparisons of businesses. (Vanderford 2021)

He notes, however, that the rule would open up firms to litigation for making mistakes in reporting, and he also discussed a

500-page proposal for a set of climate disclosure requirements that would, if adopted, be among the most expansive and complex disclosure require-ments the agency has yet put forward. . . . Observers already have noted that the new regime would require companies to expend considerable resources to craft these disclosures. (Vanderford 2021)

There is little question that adding sustainability metrics to manage-ment is complicated, and we will make mistakes as we learn how to do this. Just as financial reporting keeps accounting firms in business, com-plying with environmental sustainability metric reporting requirements will cost companies serious amounts of money and fund a growing cadre of sustainability professionals. But if we want to grow our economy with-out destroying our planet, we need to do a better job of measuring and managing our environmental impacts. I am encouraged by the fact that these disclosure and measurement issues have finally reached the politi-cal agenda.

To paraphrase a "Druckerism," or a central principle of management guru Peter Drucker's approach to management, "in order to manage something, you must be able to measure it." Without measurement, you can't tell if the actions of management are making things better or worse. We will not have adequate management of environmental sustainability without these measures. The SEC proposal is an important first step. It must proceed, but then be built on and improved.

The same process could be repeated for workplace equity and commu-nity impact—these are social and governance elements of sustainability management, but not environmental sustainability, which is the focus of this chapter and the book as a whole. Environmental performance is not more important than issues of social impact and organizational gover-nance, but it is easier to measure. A greenhouse gas in the air or lead in drinking water is a physical phenomenon that can be easily detected and measured. A company's impact on a community is measurable, but the measures are more difficult to develop and analyze. While I do not think that it makes sense to measure environmental, social, and governance (ESG) elements in a single indicator, I do believe these elements of organizational performance must be measured, and performance in these areas must be integrated into overall organizational management.

Economic indicators such as inflation, unemployment, gross domestic product (GDP), the poverty line, and consumer savings have been developed and modified by the federal government, starting in earnest with the Great Depression of the 1930s. These measures enable business to make investment decisions and government to make economic policy decisions. While there are measures of pollution and toxicity, there are no summary measures of an organization's or a location's environmental sustainability. It is easy to find measures that enable you to understand an organization's or jurisdiction's financial condition and risk, but it is impossible to make a similar judgment of an organization's or jurisdiction's environmental condition or risk. A number of nonprofit organizations have developed sustainability measures, and they are used by companies to measure sustainability. In fact, the leading organizations providing ESG measurements came together in 2020 to develop a common framework. According to the Sustainability Accounting Standards Board (SASB):

> In September 2020, five leading framework and standard-setting organizations—CDP, CDSB, GRI, IIRC and SASB—announced a shared vision for a comprehensive corporate reporting system that includes both financial accounting and sustainability disclosure, connected via integrated reporting. The joint statement outlines how existing sustainability standards and frameworks can complement generally accepted financial accounting principles (Financial GAAP). . . . As outlined in these joint statements, these five frameworks and standards offer complementary approaches. These frameworks and standards are designed for unique sets of stakeholders and are based on unique definitions of materiality. Companies can use different frameworks and standards as building blocks to develop a system of disclosure tailored to the unique needs of their stakeholders. Within this system, SASB Standards fill the need for ESG disclosure tailored to investors and other providers of financial

capital. SASB Standards are designed for communication by companies to investors about how sustainability issues impact long-term enterprise value. . . . Other sustainability-related disclosure frameworks serve their own unique purposes, and ultimately, companies must evaluate and decide which tools serve their communications objectives and meet the needs of their key stakeholders. (SASB 2022)

The effort to bring these NGOs together stemmed from "a commitment to working towards a comprehensive corporate reporting system" (World Economic Forum and Deloitte 2020). While these are all worthwhile efforts, we still need to have the U.S. government and the governments of other nations take the lead and develop a set of Generally Accepted Environmental Sustainability Metrics that would be subject to constant revision and refinement over time. This can't be left to NGOs that sell their measurement and reporting services to private, for-profit corporations. We need government to play this role and provide these measures for no charge. Organizations such as SASB, like accounting firms, can then be licensed to audit and produce reports of environmental sustainability measures. As noted earlier, metrics are needed because without measurement, you can't improve current levels of performance or know whether you are doing so. Without environmental sustainability metrics, sustainability management is not possible. I am encouraged by the SEC's proposal in early 2022 to require the disclosure of climate risk and greenhouse gas emissions by publicly traded companies. This is a good place to start, although as I indicated earlier, it is only a start.

GOVERNMENT SUPPORT THROUGH GRANTS AND TAX CREDITS FOR RESEARCH

Up to now, I have not discussed an element of climate policy that is everyone's favorite, but that I do not favor: the carbon tax. It is not difficult to understand the appeal of a carbon tax. It's an elegant, straightforward solution to the climate change crisis. You charge enough for something, and people will stop using it. The only problem with a carbon tax is that it is politically infeasible in the United States and most other places on Earth.

The math makes sense, but politics is not rational and is far from mathematical. The carbon tax is politically infeasible in France (as evidenced by the yellow vest movement) and in the United States because energy is a large and essential part of a household budget. In France the workers revolted against proposed increases in energy costs. This is because most energy consumption is nondiscretionary. A carbon tax would be like taxing food to combat obesity—it might reduce calorie intake, but its impact on poor and working people would be devastating. The argument that poor people would get a rebate, so they would not suffer from such a tax, operates better in theory than in reality. Moreover, political opposition to higher energy costs ensures that a carbon tax would rarely be enacted by any government.

The policies that we should push include increased funding for the scientific research that will make renewable energy more efficient, reliable, and cheaper. We should also continue to push for tax deductions and credits that make renewable energy less expensive to use. I'd like to brand renewable energy as the low-cost alternative to dirty, polluting fossil fuels. Let's focus on the visible, short-term negative impact of fossil fuels and price renewable energy so it always costs less than fossil fuels. Which has more political appeal: higher energy prices or lower prices? Jessica Green states that "the evidence indicates that carbon pricing has a limited impact on emissions" (Green 2021). I believe that we need to discuss modernizing the energy system in the United States to make it more efficient, reliable, and cheaper. If we focus on improving the energy system rather than solving climate change, the probability of consensus grows.

The technology of air, water, and waste management has advanced dramatically since we created the Environmental Protection Agency (EPA) back in 1970. I believe that decarbonization is in the early stages of the same process. The technology that we have now can get us started, but if it were really where it needed to be, it already would have replaced fossil fuels. Electric cars are a good example. Yes, we need more charging stations, and public policy should do even more to encourage early adoption. But what we really need is a battery that is so good that it can deliver a charge for 500 or 1,000 miles. We need an electric vehicle that costs less than today's internal combustion vehicles. Those electric vehicles will require technological innovations that I am certain we will see eventually, but they are not yet available. Those technologies will make the internal combustion engine obsolete.

The transition to renewable energy and electric vehicles has begun, but additional technological innovation and infrastructure investment will be needed if it is to succeed. The larger problem will be the greenhouse gases produced when we manufacture steel, cement, and food. These industrial processes must also reduce their production of greenhouse gases, and developing the technology needed for these changes will be a massive national undertaking. While this issue is not as important as energy decarbonization, it is important. In fact, Rebecca Dell, writing for the *New York Times*, notes:

> Last year, around the world, nearly two billion tons of steel was produced—more than 500 pounds for every person on earth. And at least 30 billion tons of concrete, or nearly 9,000 pounds for each of us. The scale can be hard to believe, until you look at a runway or a suspension bridge and contemplate what was required to build it. But all the comfort, security and convenience provided by things made of steel and concrete comes at a cost. Making steel and cement—the main ingredient in concrete—generates about 15 percent of all global emissions of carbon dioxide, the gas most responsible for the climate crisis. In the United States, industrial sources like steel mills and cement kilns are also the leading source of some of the most damaging types of air pollution. We can't solve climate or pollution problems if we don't clean up these industries. (Dell 2021)

In addition to energy and industrial decarbonization, we need to do a better job of recycling. Over the last century, we have mined more and more raw materials and manufactured products from those materials. Increased prices for mined raw materials and improved recycling technology have enabled some industries to rely more on recycled materials than ever before. A report by Bob Tita in the *Wall Street Journal* in March 2022 detailed this trend in the aluminum business:

> U.S. aluminum consumption grew by 11 percent last year, bouncing back from 2020's pandemic-influenced reduction, according to the Aluminum Association trade group. . . . To meet rising demand, aluminum companies

are doubling down on recycling, melting more scrap to increase their output of aluminum. . . . More than 40 percent of the country's aluminum supply already is produced this way, making the U.S. one of the world's biggest consumers of aluminum scrap. The U.S. is one of the world's biggest exporters of aluminum scrap, too, with 2 million metric tons of it sent overseas last year, according to government data. . . . Melting scrap for aluminum uses about 90 percent less electricity than producing aluminum in a smelter from refined bauxite ore, analysts said. (Tita 2022)

One problem with recycling is that our current system requires humans to sort garbage and relies on markets for raw materials that tend to fluctuate. One element of a solution to this issue is a more automated system of waste sorting and waste mining. Early stages of the development of this technology are now underway. According to Lori Ioannou and Magdelena Petrova:

U.S. companies and researchers are developing AI [Artificial Intelligence]-assisted robotic technology that can work with humans in processing plants and improve quality control. The goal is to have robots do a better job at sorting garbage and reduce the contamination and health hazards human workers face in recycling plants every day. Sorting trash is a dirty and dangerous job. Recycling workers are more than twice as likely as other workers to be injured on the job. . . . The profession also has high fatality rates. The way the robots work is simple. Guided by cameras and computer systems trained to recognize specific objects, the robots' arms glide over moving conveyor belts until they reach their target. Oversized tongs or fingers with sensors that are attached to the arms snag cans, glass, plastic containers, and other recyclable items out of the rubbish and place them into nearby bins. (Ioannou and Petrova 2019)

An academic study by Henning Wilts et al. published in the journal *Resources*, called "Artificial Intelligence in the Sorting of Municipal Waste as an Enabler of the Circular Economy," analyzed an effort to utilize artificial intelligence (AI) in waste sorting. It reported that "AI-enabled robots could improve working conditions for human waste management workers and improve the purity of the waste stream" (Wilts et al. 2021).

This technology is relatively new but holds enormous potential. There are a number of economic factors that contribute to the likelihood that waste management will eventually be dominated by waste mining and resource recovery. First, the cost of managing solid waste is increasing. The land used to create landfills is becoming more expensive, and in many areas, it is simply not available at all. More and more waste is being processed by some form of technology. It is obviously desirable if the expense of this technology and waste management process could be offset by the sale of resources. Second, the cost of mining raw materials is growing, as these materials must be mined from less readily accessible locations and due to the expense of complying with regulations protecting the environment near mines which are finally, if slowly, being implemented. All of this provides a *potential* economic base for recycling.

The environmental case for the circular economy is obvious. We need to end the linear model of mining, using, and discarding resources. This model damages the environment and results in ever-increasing costs. Raw materials will increase in price, and the cost of remediating toxics in the biosphere will continue to grow. The linear model may have been workable for a planet of a billion people who did not consume much, but it is unworkable on a planet of eight to ten billion humans who consume the way that many of us do. We are slowly starting to develop a circular economy where what used to be "waste" is being redefined as a resource. We have a long way to go before we end the linear economy. Where possible, we need to design consumer items for remanufacturing. Computers and other electronics should be designed so they can be disassembled for their raw materials or built to form the base for the next generation of these devices. Food and sewage waste should be used to manufacture fertilizer. Some of this resource reuse requires new product design; some requires more advanced waste sorting and waste mining facilities.

The environmentally sound mining of waste streams for resources requires the use of renewable energy. These waste-reuse processes are energy-intensive, and both the financial and environmental costs of utilizing these processes need to be minimized. That means that they need to be powered by renewable energy. While waste sorting and mining facilities will be capital intensive, the revenue streams to support these capital expenditures can be generated by selling raw materials and redirecting the funding currently allocated to waste transport and tipping fees at dumpsites.

Siting waste mining facilities will be a political nightmare. No one will want to live near these facilities. One possible solution would be to construct them on or near existing landfills, waste transfer stations, or other waste management facilities. The garbage trucks are already driving in. Better-designed facilities might include tunnels and garage structures to reduce the visibility and environmental impact of truck traffic. They should certainly include advanced emission and effluent controls. In the case of landfills, there may come a time when they too will be mined for resources. It might be a good idea to locate waste management and mining operations near them.

The issue of waste management is about as unglamorous as can be. Political leaders are reluctant to be publicly identified with waste problems or solutions. No one seems to want to cut a ribbon on a new waste management facility. And yet an environmentally sustainable city is not possible without a system for handling (and hopefully benefiting from) the reprocessing of its waste. Aluminum is one metal that has already developed a cost-effective system for reuse. Recycling the rare earth minerals used in electric vehicle batteries also has proved to be highly profitable. In fact, as electric vehicles become more common, we can expect to see them mined for raw materials when they end their useful life.

It is important to understand the way that aluminum has integrated reuse into its business model and how aluminum might lead the way for other industries. As Bob Tita reported in the *Wall Street Journal*:

> Aluminum beverage cans have some of the best recovery and reuse rates of any recycled household consumer products. About 70 percent of new cans are made from old cans, according to the Can Manufacturers Institute. . . . Scrap aluminum is becoming more important to the aluminum industry as the number of smelters, which produce virgin aluminum, has been shrinking for years. Six smelters continue to operate in the U.S., down from about two dozen 20 years ago, as executives have said aging equipment and rising electricity costs make them increasingly expensive to operate. . . . Rising scrap prices are providing more incentives for collecting scrap and investing in the equipment to process it. Used beverage

cans are trading at $1.38 a pound, up 78 percent in the past year, according to S&P Global Commodity Insights. The price of old aluminum sheet scrap is 38 percent higher from a year ago. (Tita 2021)

Aluminum reuse is not typical of most metals today, but we can expect the same economic logic that worked for aluminum to become common for other materials in coming years. New technology and enlightened leadership will be needed before we will see the widespread construction of municipal waste management and material mining facilities. Nevertheless, the construction of such facilities is essential for the development of a circular economy, and the evolution toward a circular economy is a prerequisite for sustainable cities. AI and automation, along with revenues from the sales of recycled materials, should make these facilities financially viable in the near future.

Another key area of government investment, central to a green economy transition, is basic and applied research. While environmental advocates claim that the current technology in green energy and production is adequate for the transition to a renewable resource economy, it is not. If it were, it would be replacing existing technology now. My view is based on a historical analysis of technology adoption and replacement. A new technology replaces an old one when it is less expensive, more reliable, and better (i.e., includes more desirable features) than the old one. Smartphones replaced cell phones, which replaced land lines. Refrigerators replaced ice boxes. Electric lights replaced candles and gas lights. Autos replaced horse-and-buggy rigs, and the electric vehicle is about to displace the internal combustion engine. Containerized shipping and bar codes replaced stevedores with paper freight manifests. Cloud computing on the internet replaced multiple local mainframes. GPS has replaced paper maps. Wireless internet has replaced most wired internet. In each case, the new technology replaced the old because it had features and/or price advantages that the old technology lacked.

As of now, solar cells and windmills have not yet replaced fossil fuels for generating electricity. We need less expensive solar cells and batteries to displace fossil fuels. The commercialization of technology to do this will make someone very wealthy, but the basic research needed for this transformative technology will probably not come from corporate research and development, but rather from government-funded basic and applied research. The payoff of any one technology is too uncertain to ensure a reasonable chance of a return on equity. In the United States, our research universities, national laboratories, and defense-funded basic research institutions have created path-breaking technology, including microcomputers, cell phones, jet engines, GPS, and the internet itself. Today's renewable energy batteries are better than they used to be, but they are still too expensive, too inefficient, and too dependent on rare earth metals and toxic materials. Solar cells have similar deficits. But these technologies are better now than a decade ago, and they can and will be improved. Government funding of basic and applied research will accelerate the commercialization of needed green technologies.

GRANTS AND TAX INCENTIVES FOR UTILITIES

The idea is simple enough: The government invests in green infrastructure and stimulates the private investment needed to decarbonize and modernize the American economy. There is big money in green infrastructure in both blue states and red states. And a policy to invest in modernizing the electric grid, subsidizing renewable energy, promoting electric vehicles, and making our homes and businesses more energy efficient can be a political winner. Modern high-capacity transmission lines can enhance the grid's flexibility. Powerful economic interests and labor unions will support infrastructure investment. The heavy lift will be raising the tax rates on the wealthy to pay for it. Of course, we don't need to raise the rates overall—just impose a minimum tax on all the very wealthy people who manage to pay little or no tax. Americans have gotten used to paying low taxes and living in debt, and to invest in a new energy system we will need to invest significant capital. I discuss the specifics of energy infrastructure in greater detail in the following chapter.

Tying economic development to environmental protection is a winning political strategy. Instead of focusing on enacting regulations and taxes that punish polluting behavior, we invest in infrastructure and tax incentives that promote environmental sustainability. The goal is to lower the cost of renewable energy rather than raise the cost of fossil fuels. Another is to develop the technology needed to sort waste and make our waste stream part of our material stream. These investments connect environmental protection to employment. I have long believed that the only practical approach to climate policy is to encourage the rapid development and implementation of renewable energy technology. Energy use is too embedded in our daily economic life to reduce its use beyond the considerable gains that we could achieve via energy efficiency.

In the case of fossil fuels and the power grid, it is not simply that the current energy system damages the environment and contributes to climate change; it is that the system is ripe for modernization. Centralized power generation and distribution are vulnerable to interruption from extreme weather, terrorism, or computer hacking. It is capital intensive and, as a regulated monopoly, open to political influence. As we learned during Russia's invasion of Ukraine in 2022 and the Arab oil embargo of the 1970s, fossil fuel dependence distorts foreign policy and requires democracies to provide financial support to autocracies. Distributed, decentralized, and renewable energy sources are less vulnerable to mass failure, potentially less damaging to the environment, and as technology develops, likely to come down in price as well.

Federal funding to modernize the energy system and make it more efficient and green may provide the resources to invest in a lower-cost and more reliable energy system. The United States will need such an energy system to remain competitive in the global economy. Its electric grid is increasingly unreliable. According to the U.S. Energy Information Agency: "On average, U.S. electricity customers experienced just over eight hours of electric power interruptions in 2020, the most since we began collecting electricity reliability data in 2013" (Lindstrom and Hoff 2021). Most interruptions are caused by major weather events. Climate change is responsible for more frequent and intense weather events. Without redesign and reconstruction, we should expect increased energy blackouts.

That is the sales point that seems to be missing when we discuss the energy system and climate change. Even if fossil fuels were not destroying the planet, they remain a technology ripe for displacement. In the long run, fossil fuels will be far more costly than renewables. Fossil fuels are finite, and while there is plenty of supply left, it's getting harder to get to it. Some of the limits are physical, and some are political. The Sun, by contrast, will last longer than our species. The technology of solar cells, batteries, and wind power continue to improve and get more efficient and less expensive—sort of like computers and smartphones. A multimillion-dollar computer in the 1970s had far less computing power than today's $300 smartphone. The source fuel for renewables is free. Contrast that to fossil fuels. Oil, coal, and gas must be extracted from the ground at a cost to the pocketbook and ecosystems, transported to where they are burned (at more cost), and then finally burned—expense on top of expense. It's a technology that is being disrupted and displaced by renewable energy.

For the foreseeable future, we still need an energy grid, and with high-capacity transmission lines, we might transmit renewable energy from sunny and windy places to cities. But efforts like the ridiculous one in California to tax homes with solar arrays to pay for the grid will not work, and grid finance is going to be more difficult when distributed power generation reaches an as-yet-unknown tipping point. We can expect to see electric utilities and their regulators pushing back on efforts to promote renewable energy. A recent proposal in California to tax solar arrays was lampooned by former governor Arnold Schwarzenegger in the *New York Times*. According to Schwarzenegger, the state was proposing

> a new monthly "grid participation charge" that would average an estimated $57 a month for solar customers. People who power their homes with fossil fuels wouldn't pay this. So let's call it what it is: a solar tax. This solar tax would also apply to customers who invested in batteries to store that solar energy. Battery storage is critical for the transition to clean energy and grid resilience. But this tax will only discourage that progression. Moreover, the commission would cut credits to new solar customers (and some older ones) as much as 80 percent for the electricity they don't use and send to the grid under the net metering program. Those credits in turn can lower their utility bills. This is just another case of the big guys—the investor-owned utilities—fighting for themselves and hurting people who have invested or want to invest in solar panels. (Schwarzenegger 2022)

The political reaction against the California Utility Commission's effort in 2022 was swift. As a result, Governor Gavin Newsom and the commission soon backed off.

The institutional and financial interests invested in fossil fuels and the electric grid are major obstacles to modernizing our energy system. While current owners of solar arrays and batteries often sell back their excess to the grid, thus lowering the cost of energy on the grid, an effort to tax household solar could result in decisions to disconnect from the grid entirely. Technology may develop that will make cutting the energy cord less risky and more common, leaving those dependent on the grid with higher costs. We have seen this with telephone landlines and cable television. Why should electricity be immune from similar influences?

The issue in California was that the payment to homeowners and businesses selling energy back to the grid may have been set too high in an effort to encourage renewable energy investment. The cost of maintaining and updating the grid needs to be paid by someone. The capital investments needed are partly subsidized by the federal infrastructure climate provisions. But the long-term shape and financing of the electric grid will need to be rethought and refinanced as part of the effort to modernize and decarbonize the energy system. Utilities are starting to respond to changing conditions and are increasing their capital spending. While utility investment will increase our power bills for a while, they are critical to decarbonization. According to Katherine Blunt in the *Wall Street Journal*:

> American utilities are planning their biggest spending increases in decades to upgrade aging grids, prepare for electric vehicles and make the transition to renewable energy—moves poised to further boost power costs as consumers face historic inflation. The plans propose tens of billions of dollars in spending in the coming years to reduce carbon emissions, partly in response to state and federal mandates, and to replace aging infrastructure that has become more prone to failure. Edison Electric Institute, an industry trade group, expects that utilities will invest roughly $140 billion each year in 2022 and 2023, substantially more than any year since 2000, when the group began tracking spending. Executives said the investments are critical to meeting renewable-energy targets and bolstering the reliability of the grid as outages become longer and more frequent. Climate change, they said, has heightened the need to simultaneously hasten the shift to carbon-free electricity sources and upgrade the grid to withstand severe weather patterns scientists link to rising temperatures. (Blunt 2022)

The investment is in improving grid reliability and meeting carbon emission limits set by government. While not acknowledged in Blunt's piece, I also suspect that the utilities see the cost curves and know that renewable energy is getting less and less expensive.

Since I live in an apartment in New York City, I have no place to put rooftop solar. But a number of companies are working on placing solar cells in window glass. That technology shows enormous promise and should lead to a time when people in multifamily homes or offices in skyscrapers will be able to generate their own electricity.

While all these technological developments will help reduce greenhouse gases, they also promise lower costs and more reliable electricity when coupled with advancing battery technology. At one time, homes were heated by fireplaces and light came from oil or gas-fueled lamps and candles. Those technologies were displaced by oil, gas, and electric heat, as well as electric light bulbs. There is no reason to believe that the current method that we use to power our homes will continue indefinitely. In the future, electric utilities will play a different role than they currently play in powering our economy. We all have an interest in an energy transition that prevents the bankruptcy of electric utilities and encourages their active participation in the transition to renewable energy. We should not use taxes on renewable energy to fund the transition, at least until the use of renewable energy is more widespread. California has over a million installations, but it is a state of over thirty-nine million people. They have a long way to go, and the rest of the country will take an even longer road to renewable energy.

It will take national-level resources to provide the capital needed to modernize the grid. By relieving utilities and their ratepayers of some of these capital costs, utilities can reduce costs by making the grid less wasteful and less dependent on fossil fuels. A large-scale, national project like modernizing the electric grid will require subsidies, but once it is complete, some of the continuing costs of grid maintenance and energy generation can be reduced and designed to accommodate a different load than the current system seeks to meet. In short, grid modernization should act like most forms of infrastructure investment: Eventually, it pays for itself in economic benefits.

The transition will be complicated, and financing needs will vary by location. Interest group politics will feature intense lobbying by utilities, fossil fuel companies, renewable energy companies, and environmentalists. But the goal should be a lower-cost, more reliable, and less polluting energy system. By welding those three elements together under the umbrella of modernizing our energy system, we can create a broader coalition than an effort dominated by the goal of greenhouse gas reduction. Energy modernization is the goal, but greenhouse gas reduction is a much-needed by-product.

BUILDING REGULATIONS THAT ENCOURAGE AND REWARD RENEWABLE RESOURCES, RECYCLING, SHARING, AND MITIGATING ENVIRONMENTAL IMPACTS

While I am stressing the need for investment and incentives, we still need environmental rules and standards. We need to define and measure pollution if we are to reduce and prevent it. I do not favor rules because I like them, but because in a complex, interconnected world, we need them.

Let's go back to the basics: No one likes being told what to do. Watch New Yorkers at a crosswalk, ignoring the "Don't Walk" sign, and you'll see what I mean. But we live in a more crowded, complex, and interdependent world, and our health, safety, and welfare depend on effective regulation. The attack on regulation is about the balance between freedom and restraint. Effective regulation seeks to balance these goals. Sophisticated regulation requires change and adjustment, and systems of regulation in the United States can easily get stuck in place. Conservative opposition to regulation has been going on in this country for many decades, having picked up momentum when Ronald Reagan was president in the 1980s. Regulations can and should be improved, but rules are the price we pay for the benefits of jet travel, innovative construction materials, the internet, the smartphone, and energy and water delivered to our homes as if by magic.

Technology influences our behavior, and those behaviors can cause us to harm ourselves and each other. One example is distracted driving. Highway accident rates, auto insurance rates, and deaths are increasing as drivers look away from the road to view texts, maps, and who knows what else on their smartphones and dashboards. If the only victims of distracted driving were the distracted drivers, it might not require action by government, but those of us sharing the road with these folks are in danger too. Rules and regulations are needed to curtail the distracted driver's freedom to text while driving. Someone needs to enact these laws, provide the details of rules to be followed, develop new safety technology, and ensure enforcement to reduce this danger.

From the beginning of the Industrial Age to the 1970s, air pollution from factories, power plants, and vehicles grew with the growth of the U.S. economy. But by 1980, we learned that we could increase our use of material goods, energy, and vehicles while reducing pollution. "All" we needed to do was require companies to use technology and best management practices to control pollution. American environmental regulations have been remarkably effective. And while some pollution was exported, most sources remained and improved their environmental performance. Since 1980, the absolute level of pollution in the United States has decreased. According to the EPA, from 1980 to 2015 the U.S. GDP grew by 153 percent, our population grew by 41 percent, and vehicle miles traveled grew by 106 percent, but air pollution declined by 65 percent (EPA n.d.). That is the definition of successful regulation. China and India would love to be able to make the same claims, but they have not yet implemented effective environmental regulation. U.S. water pollution and toxic waste releases have also been reduced over the past several decades. With the antiregulation fever accelerating in conservative-led states, we had better be careful to protect the basic structure and fabric of our largely successful environmental laws.

The image of freedom, of individuals and businesses doing what they want without fear of restriction or taxes, is powerful and seductive. But we don't live in log cabins surrounded by fields or forests. We live on top of each other or right next to each other, and we move around in congested planes, trains, and automobiles. Our actions affect each other. We need rules to motivate constructive, socially responsible behavior. We need sufficient enforcement to ensure that people believe that there might be sanctions if they do not follow the rules. And to keep the rules up to date, we need to understand technology's changing influence on our behavior and on our planet.

Rules on texting while driving wouldn't have been developed if we hadn't invented cars, smartphones, and text messaging. Both technology and the population are growing. Global warming would be less challenging on the planet of three billion people that we had fifty years ago than it is on the planet of over eight billion people we have today. It should not be surprising that governments at all levels and in all places are working to develop the capacity to manage the growing economic interactions of a growing population.

While its management could always be improved, the EPA is one of the leaner federal agencies. Most of its staff work out of its ten regional offices, and the agency's growth has been quite slow over the last several decades. In 1980, there were 13,000 staff at the EPA; that number grew to 16,300 in 1990 and 17,700 in 2000. But it dropped to around 17,300 in 2010 and then to about 15,400 in 2016. In fact, the agency shrunk by 1,600 employees during President Barack Obama's time in office. I would argue that one reason for some of the EPA's inflexibility on green pollution control has been that its lack of staff and other resources did not allow it to develop the capacity needed to apply standards with the necessary flexibility.

One of the EPA's responses to its small staff and budget has been to delegate most regulatory oversight to state governments. State and local law and regulatory capacity has grown, even as the EPA's has not. But the agency's oversight of state and local activity has often been rigid and unimaginative. We saw that in New York during an effort to meet water quality standards with green infrastructure instead of traditional concrete and steel water pollution control construction. The EPA was not confident that planting vegetation and relying on natural filtration processes could reduce the impact of nonpoint sources of water pollution or absorb the amount of rainwater needed to reduce combined sewage overflow. It took local and state officials several years to demonstrate the cost-effectiveness of these innovations and to get federal officials to agree to them. A larger EPA with a larger science budget could have moved more rapidly to conduct the studies needed to verify state and local claims.

The green infrastructure battle was over means rather than ends. Federal, state, and local officials were all working hard to meet national water pollution control standards. Businesses and government knew that these standards must be met. The danger of an antiregulatory federal government is that the probability of enforcement will be reduced in some states. While all the states have needed to develop the legal and administrative capacity to implement federal environmental rules, some states take these rules more seriously than others. The threat of federal oversight provides only minimal quality control. While Michigan and EPA Region V dropped the ball on overseeing and addressing the lead-contaminated water supply in Flint, there are many other examples of effective national regulatory oversight. In New York, California, and probably a dozen other states, federal quality control is not needed; local support for environmental protection is strong enough to ensure that national standards will be taken seriously.

The idea that environmental regulation costs jobs and depresses economic growth is a fallacy. Certainly, there are examples of businesses harmed by environmental regulation. But overall, the new technologies, new methods of production, and reductions of illness caused by environmental regulation provide far more economic benefits than costs. They also make our economy more efficient. For example, fuel efficiency standards encourage the development of new technologies and vehicles that use less fuel. Firms that buy these new, more efficient vehicles reduce their cost of doing business.

Some business leaders believe that government regulation is rigid and unthinking and inhibits economic growth. There are plenty of examples of regulation that is counterproductive. Some advocates of government regulation believe that these business leaders are blind to the benefits of regulation due to ideology. Both sides are right but miss the point. In a complex world with growing numbers of interactions between businesses and their supply chains and consumers, the potential for danger and malfeasance continues to grow. When regulation fails, our litigious society can simply resort to civil suits as a method of regulating behavior—a method that is as blunt as it is expensive. Well-thought-out rules, intelligently applied and judiciously enforced, could provide business with a clear guide to the rules of the road. On the other hand, the absence of rules provides an invitation to anarchy, danger, and declining economic confidence. We have gotten used to regulation and antiregulation as a set of on and off switches governed by so-called liberal and conservative ideologies. We should be thinking about regulation as a way of ensuring that economic activity can proceed with the least possible amount of unanticipated negative impacts. *We don't need more regulation or less regulation, but more effective regulation.* This requires all stakeholders to work together to create reasonable rules that are realistic and implementable, but also improve outcomes.

In a *New York Times* piece on business attitudes toward regulation during the Donald Trump administration, Binyamin Appelbaum and Jim Tankersley observed:

> There is little historical evidence tying regulation levels to growth. Regulatory proponents say, in fact, that those rules can have positive economic effects in the long run, saving companies from violations that could cost them both financially and reputationally. . . . But in the administration and across the business community, there is a perception that years of increased environmental, financial and other regulatory oversight by the Obama administration dampened investment and job creation—and that Mr. Trump's more hands-off approach has unleashed the "animal spirits" of companies that had hoarded cash after the recession of 2008. (Appelbaum and Tankersley 2018)

These macho management concepts of "animal spirits" are as ridiculous as they are popular. There is little actual evidence that fear of regulation is a major component of business decision-making. No one likes government telling them what to do, but stable and predictable rules create the level playing field required for investment and fair competition. Still, the antiregulatory folks have a point. Regulation is often rigid, unintelligently applied, and slow to change even as conditions do. It can lead to dysfunctional behaviors by businesses that benefit no one—not even them. State and local governments are also regulated by the federal government, and they too can find federal rules mind-numbing and counterproductive. But the problem is not the idea of regulation; it is the effectiveness and efficiency of regulation.

The analogy I would draw is one pointed out by David Osborne and Ted Gaebler in their 1993 book *Reinventing Government*. Osborne and

Gaebler were responding to decades of antigovernment ideology starting with Barry Goldwater's presidential campaign in 1964 and culminating in Ronald Reagan's election in 1980. Reagan famously declared that government not only couldn't solve society's problems, but it was the problem. *Reinventing Government* argued that government had a key role to play in national life, but inadequate government management resulted in ineffective and inefficient government programs. The answer was not to reduce government but improve it. President Bill Clinton appointed Vice President Al Gore to chair a National Performance Review in an attempt to take the ideas in *Reinventing Government* and apply them to the federal government. My own view is that these ideas had little impact on the federal government, but they did have a much bigger impact on state and local governments—especially local. Local governments deliver most of the services that the public sees, and the pressure to improve performance is unrelenting. The result has been decades of improved service, in part based on the ideas of Osborne and Gaebler.

There has been no similar attempt to reinvent regulation. It remains dominated by formal legal thinking and is one of the least creative and least innovative functions performed by the federal government. Improving regulation requires that the federal government utilize some of the enhancements that modern management has brought to other production functions. Regulators must increase their use of strategic thinking, communications and information technology, and marketing (especially using social media). Regulators need to do a better job of understanding the views and experiences of stakeholders—the regulated community, advocates of regulation, and the broad public. I've been pushing for changes to regulatory strategy for decades, and in 2005, I wrote a book on that topic with my colleagues Sheldon Kamieniecki and Matthew Cahn, entitled *Strategic Planning in Environmental Regulation*. In it, we argued that regulation was too often defined as command-and-control standards and enforcement, and instead government should consider a wider range of methods to influence private behavior to meet public needs. Tax incentives, education, and a wide range of coercive and noncoercive methods are available to influence behavior, and, in this day of endless information and inexpensive communication, regulations can be individually tailored to maximize benefits while minimizing costs. The web, drones, social media, sensors, automation, computer controls, focus groups, and social psychology can be brought into a world still characterized by random inspections, paper documents, lawsuits, threats, counterthreats, conflict, and snail mail.

Unfortunately, since we are stuck in a useless debate between pro-regulation and antiregulation advocates, we are missing opportunities to improve regulation. Regulatory agencies like the EPA have been attacked and cut for decades. The conservative strategy of "starving the (governmental) beast" has ensured that there are no resources available to improve regulation and update it for the twenty-first century. There may have been some support for additional regulation during the Obama years, but there was no discussion of reinventing regulation and figuring out a way to create a more agile and flexible way to influence private behavior to serve the public's interest. We need to reengineer regulations; we do not need to increase or destroy them.

It's time to turn the page and reinvent the whole concept of regulation. Regulation is not cost free, but pollution, poisoned drugs and food, unsafe work sites, and unfair labor practices bring high costs too. There are very few advocates for poisoned food and dangerous work sites. The issue is who pays for the cost of compliance and how and when those costs are charged. This silliness of cutting the number of rules or repairing the damaged psyches of discouraged business leaders really needs to be discarded. We need to grow up and figure out how to grow economic activity in ways that preserve the planet, protect workers, promote fairness, and protect our children's future.

The United States is not the simple, uncrowded place its founders knew in 1776. Technology has transformed economic life and freed us from a great deal of drudgery and pain. But it also has introduced dangers that were unimaginable two and a half centuries ago. Terror, mass financial fraud and manipulation, computer hacking, disinformation, nuclear war, biological contamination, toxic chemicals, invasive species, climate change, and species extinction are real and present dangers. Russia's brutal invasion of Ukraine in 2022 brought home to Europeans and many others the point that there is no guarantee that our comfortable way of life will continue. COVID-19, western forest fires, and the endless stream of hurricanes are all examples of the impact of the natural world on our way of life. We are all one disaster away from being displaced refugees. To maintain and improve our way of life, we need to police ourselves and make sure that the rules of correct behavior are clear, fair, well understood, and creatively applied.

Our goal should be to maximize individual freedom, but to balance that freedom against the needs of our community. The goal of regulation is not to show who's in charge or to declare winners and losers; it's to manage the complex interactions that provide us with the benefits of our technological age at the least possible cost. Regulation is simply too important to leave to ideologues. Our goal should be to develop a high-throughput economy that preserves the planet, while enabling the entire world to live as we do in the developed world. To do this, we need to develop both a deeper understanding of our planet and the technology to deliver material goods without destroying the environment. Antiregulatory ideology is an obstacle to the development of the technology needed to transition to environmental sustainability.

PROMOTING RENEWABLE ENERGY INSTEAD OF OPPOSING FOSSIL FUELS

Instead of taxing fossil fuels, we should subsidize and invest in renewable energy. Our goal is to lower the price of energy and make our energy system more efficient, equitable, and less damaging to the environment. Let technology and the market get rid of fossil fuels, and use politics and public policy to accelerate the development and adoption of renewable energy technology and to modernize our decaying electrical grid.

The only politically feasible way of combating climate change is to invest in research and development and infrastructure to make our energy system more efficient and less dependent on fossil fuels. The goal is to build a more efficient, modern, and lower-cost energy system. We need a system that is based on lower-cost renewable energy, widespread energy storage, and distributed generation of energy, and uses energy as efficiently as possible.

Energy utilities all over the United States are encouraging their customers to adopt energy efficiency measures. Some states, like California and New York, tax energy use to generate funds that are invested
in energy efficiency. A climate policy that focuses on energy efficiency
is already in place in many places and is politically popular. A similar
approach to renewable energy-one that emphasizes technological innovation designed to lower prices while maintaining reliability-can drive
fossil fuels from the marketplace and ease the climate crisis confronting
our planet.

The American states that modernize first will end up with a more-
reliable and lower-cost energy system. As states commit to greenhouse
gas reduction goals, assuming that they meet those targets, they will be
replacing fossil fuels with renewables, but also investing in microgrids
and other technologies that will make energy use more efficient and
reliable, ranging from light bulbs to compressors, and heat pumps to
building and window insulation. The low-hanging fruit of greenhouse
gas reduction is energy efficiency, and a more efficient energy system
lowers the costs of energy that consumers have to pay. Renewable
energy is already less expensive than fossil fuels, and this price differ-
ential will probably grow. The modernization of our energy system will
make decarbonized states more rather than less economically competitive. Hopefully, fossil fuel states will recognize the bottom line and move
away from fossil fuels.

Carbon taxes and carbon markets have a role in greenhouse gas reduction because some of the reductions in greenhouse gases will come from
rules and policies mandating or encouraging reductions of carbon dioxide.
But the overall approach of raising the price of fossil fuels always clashes
with political pressure to keep energy prices down. Waxman-Markey's
climate regulation cap-and-trade bill passed the House of Representatives
in 2009 by 219–212 votes, but died in the Senate due to the certainty of
a filibuster. The failure of Waxman-Markey's cap-and-trade policy during the Obama years, and the failure of efforts to punish utilities for not
decarbonizing during the early days of the Biden administration provide
a graphic illustration of the limits of pricing carbon. Subsidies that help
reduce the price of renewables and lower energy costs may be harder to
pay for, but they are far easier to sell as a policy.

END COMMUNICATIONS THAT SHAME CONSUMPTION
AND ATTACK BUSINESSES

Attacking business does not speed the transition to a sustainable economy. We should be building consensus and cooperation along with mutual gain, not declaring each other enemies and trying to win for the sake of winning. The sustainable cities that most of us will live in will be exciting, dynamic, healthy, and hopefully fair and equitable places. Its achievement is a goal that all of us should aspire to. Our focus should be on the development and implementation of renewable energy technologies. A nonideological consensus could be created to meet the goal of "modernizing America's energy system." Pushing for more-reliable, lower-priced, and less-polluting energy has a higher probability of success than directly attacking fossil fuel companies. If the issue is framed as banning fossil fuels from the developing world, self-interest, national sovereignty, and the pride of developing nations will prevent any international agreement. If the issue is framed as the developed world providing incentives to developing nations to build a carbon-free energy system, there is a good chance that we could move in that direction.

I believe that an attitude of moral superiority is particularly destructive to any effort to build the political support needed for making systemic change. Much of the work of building environmental sustainability requires the development of systems that enable us to live our lives as we wish while damaging the planet as little as possible. Large-scale institutions are needed to manage sewage treatment, waste reuse, and drinking water; develop renewable energy; and build a modern energy grid. Government policy is needed to ensure the conservation of forests, oceans, and biodiversity. Pandemic avoidance requires global, national, and local systems of public health. Climate change mitigation and adaptation also require collective action. What, then, can individuals do?

As individuals, we make choices about our own activities, and they inevitably involve choices about resource consumption. I see little value in criticizing a person who flies to global climate conferences via a private or any other plane. But I see great value in considering the importance of the person's attendance at the conference and asking if the trip is an indulgence or if he or she will have an important opportunity to learn, engage, and teach this topic. The COVID-19 pandemic taught us how to attend events virtually. There is little question that live presence at an event enables a type of communication that can't be achieved virtually. Many times, you will determine that the financial and environmental costs of the trip are far outweighed by the benefits. Those are the times when you should travel. My argument here is that the thought process, the analysis of environmental costs and benefits, lie at the heart of an individual's responsibility for environmental sustainability. Individuals need to think about their impact on the environment and, when possible, minimize the damage that they do to the planet.

Everyone needs to turn on the lights at night, start the shower in the morning, turn on the air conditioning in hot weather, and possibly drive somewhere on Mother's Day. I would never argue that you should give up these forms of consumption. Instead, I believe that we should all pay attention to the resources that we use and the impact that that use has. We are responsible for that thought process and the related analysis of how we, as individuals, might accomplish the same ends with less environmentally damaging means.

I consider individual responsibility and the thought process and value shift that stimulate individual action as the foundation of the social learning process required for effective collective action. In other words, individual change and collective system-level change are interconnected. The fact is that on a planet of eight billion people, getting back to the land is not feasible. We keep making more people, but we are not creating (nor can we create) much new land. There is an absolute limit to our ability as individuals to reduce our impact on the planet. Therefore, system-level change is absolutely needed. But this change requires individuals to understand the need for it, along with a well-understood definition of the problem. The cognitive dissonance of identifying a problem but never acting on it is difficult to live with. If you see a poor child on the street begging for food, you can provide that child with food and money while continuing to support public policies that address the child poverty issue at the system level. In fact, the emotional impact of that child's face may well provide the drive that leads you to fight harder for policies that would prevent that child from needing to beg in the first place. We learn by example, and vivid experiences and cases can lead to transformative systemic change.

While I consider individual and collective responsibility to be connected, without collective systems and infrastructure supporting environmental sustainability, there are distinct limits to what individual action can achieve. That is why I see no value in shaming individuals for consuming fossil fuels, eating meat, or buying a child a Mylar birthday balloon.

As my mentor, Professor Lester Milbrath, often argued, the only way to save the planet is through social learning that would enable us to "learn our way to a sustainable society." He made this argument in his path-breaking work: *Envisioning a Sustainable Society: Learning Our Way Out* (Milbrath 1989). In his view, the key was to understand environmental perceptions and values and to build on those values and perceptions to change both individual behavior and the institutions their politics generated. To Milbrath, the human effort to dominate nature was working too well, and a new approach was needed. As he observed:

> Learning how to reason together about values is crucial to saving our species. As a society we have to learn better how to learn, I call it social learning; it is the dynamic for change that could lead us to a new kind of society that will not destroy itself from its own excess. (Milbrath 1989)

My view is that one method to pursue social learning is learning by doing—in other words, encouraging each of us to adopt the individual behaviors that will reduce our environmental impact. Those behaviors remind us to think about the planet's well-being along with our own. They reinforce and remind us, and as they become habit, they affect our values and our shared understanding of how the world works.

There is, therefore, no trade-off between individual and collective responsibility for protecting the environment unless we insist on creating one. In addition, in a world of extreme levels of income inequality, wealthy people who have given up eating meat have the resources to consume alternative sources of nourishment. They do not occupy the moral high ground if they criticize impoverished parents who proudly serve meat to their hungry child—quite the opposite. In our complex world, we should mistrust simple answers and instead work hard to understand the varied cultures, values, and perceptions that can contribute to the transition to an environmentally sustainable global economy. The path to environmental sustainability is long and winding and will require decades of listening and learning from each other.

GREEN TECHNOLOGIES CAN PROVIDE THE SAME BENEFITS AS THOSE THAT POLLUTE

Renewable energy provides the same electricity that fossil fuels can. The Ford Lightning 150 pickup truck introduced in 2021 has more features than its fossil fuel–based predecessor. It was so popular that Ford's initial list to reserve the truck had to be closed when it reached 200,000 customers. Consumers understood that this electric vehicle would be more reliable and have fewer moving parts than traditional vehicles. Likewise, streaming video is more convenient than videotapes and compact discs. A solar array and battery can provide the same electricity at a lower price than the grid can. Moreover, when power lines go down in a storm, your solar battery can continue to power your home.

Some green products will be indistinguishable from their more-polluting predecessors. Food grown in a field that is using AI, automation, and high-powered visualization and analytic tools will use less water, pesticides, and fertilizer but taste the same. Moreover, because they are grown with fewer resources, the food may be less expensive. If an electric vehicle battery is made from recycled rather than freshly mined minerals, the driver of the vehicle will probably be unaware of the sources behind it.

The point here is that the transition to a green economy does not mean giving up benefits. The technologies of the future will be less resource consumptive but provide similar (if not greater) benefits. The theme of sustainability as sacrifice has been in some respects a point of agreement between proenvironment and antienvironment advocates. A realistic strategy for environmental sustainability should emphasize its benefits, not its costs.

Moreover, history tells us that predictions of gloom and doom are not always accurate. A trend line can be altered by new technologies, inspired leadership, and cultural and social change. The disasters that arrive are not always the ones you predict. Putin invaded Ukraine and President Volodymyr Zelenskyy, a former comedian, rallied his country in resistance, with surprising success. No one predicted either occurrence. The population explosion feared in the 1970s came, but then it ended as a function of economic development. We learned that when children became an economic cost rather than asset, the birth rate shrank below replacement level.

We have a great need to innovate to learn to maintain and expand the comfort of our way of life. We need to develop new technologies to enable this to happen. Many of those technologies are in development or already in place. How fast they develop and the business models that will succeed are impossible to predict. But it is possible to predict that these new technologies will develop, and we won't need to sit alone in the dark with a candle in order to achieve environmental sustainability.

THE CHANGING NATURE OF CONSUMPTION AND ENVIRONMENTAL SUSTAINABILITY

As noted earlier, in the United States, 80 percent of the U.S. GDP is now in the service economy, and a growing number of people are disconnected from the part of the economy that makes things. On the consumption side, the race to have the most stuff seems to have crested for many people. Mall shopping as recreation has been reduced and replaced by e-commerce, which can be less time consuming and often is less pursued as a social activity. Malls themselves are reducing the space devoted to retail and adding space for services, offices, and even residences. When thinking about consumption, I think about lifestyle and ask two questions: (1) What do people spend their time doing? (2) How is that changing? Technology has fundamentally changed what we do with our time, and the most significant change over the past generation has been the amount of time that we are communicating with each other, the time that we spend disseminating and processing information, and the hours that we devote to being entertained.

The cost of using the telephone has been reduced dramatically since the development of the cell phone, internet, and satellite communications. Fifty years ago, an international phone call was a luxury; today, the cost of a phone call or a Skype, Zoom, or FaceTime call is so low as to be virtually negligible. Therefore, people spend hours a day communicating with friends, customers, colleagues, and family all over the world. When you add texting, email, and social media to the mix, the amount of time we spend communicating continues to grow. In the past, you assumed that if you saw someone walking alone down the street in New York City and they were talking, they were at a minimum a little weird. Today, you look for their airbuds or headphones and realize they are engaged in a phone conversation. From a waste production perspective, this form of consumption-communication has very little environmental impact. Therefore, even though it adds to the GDP it is a lifestyle change that has reduced the damage that individual acts of consumption have on the planet.

More and more people are spending their time engrossed in whatever is happening on their screens: You might be playing an interactive game with friends in different faraway locations; perhaps you're watching a movie or sports event; you might be engaged in a video conversation with a relative or friend; or you might even be reading a blog or a book. To the extent that this form of interaction replaces face-to-face human interaction in three dimensions, it is a little scary to those of us born before the internet age.

But as much as people spend time in the virtual world, there still seems to be a type of communication and interaction that requires that we sit in the same room and engage with each other. That is probably why businesspeople travel thousands of miles to have dinner with a client or a colleague. It's why, in the days when pandemic restrictions were lifted in the spring of 2022, my students were so delighted to see each other in classes, on campus, and at events. It's also probably why online forms of education will never replace (but can augment) live classrooms. Very often, electronic communication stimulates in-person contact. But time spent in the virtual world is one that has little negative impact on the real world of ecology and life.

Our changing lifestyle in the developed world has a great deal to do with the changing nature of work. Throughout human history, most people spent most of their time obtaining food, clothing, and shelter. Those activities were strenuous and labor-intensive. As noted earlier, first, humans figured out how to replace some human labor with domesticated animals, and then we figured out how to replace animals with machines. Increasingly, the production of the material items that we consume has become automated and requires less and less human labor. At the start of the twentieth century, 40 percent of Americans worked in agriculture. In 2021, it was around 1 percent. We grow more food than ever in the United States, but we grow more of it with machines than people.

While the trend began before the pandemic, we learned that work could follow us home, and that in some professions, home need no longer be close to a workplace. This led some people to build home offices, and others to reduce the number of days that they commute to work. The environmental impact of this is mixed. While energy may have been saved via less commuting, an office of fifty people cooled, lit, and heated in a central location used less energy than fifty home offices. Nevertheless, the role of place in work has probably changed for the foreseeable future, and its environmental impact requires additional study.

The transition to a service-oriented brain-based economy means that most of our work is either creative, analytic, or serving customers in some way. This has contributed to a different style of consumption when we are not at work. For some people, the absence of physical exertion at work leads them to spend recreation time working out at the gym, hiking, jogging, or biking. For others, it leads them to want to engage their brains in activities under their own control rather than for the organization they work for. Those are the folks playing Words with Friends or Minecraft sitting next to you on the subway or at the airport gate.

The other growing form of recreation is travel and tourism. People can see images of famous sites from around the world on the web, and they want to visit those places. They want to taste different foods and get a sense of how people live in different places. They want to climb spectacular mountains, sit in famous theaters and religious shrines, and be able to imagine what it would be like to live a different life in a different place. It's true that this form of consumption can be very resource-intensive and generate greenhouse gas pollution, along with waste generated by over-consumption from people in a new environment, with no work to do and all the free time in the world. It's also true that during the COVID-19 pandemic, global travel was significantly disrupted by national restrictions. If anything, I believe that COVID made people even more interested in travel. They realized how much they missed their freedom of movement and were eager to return to the road and the air.

I think that these forms of consumption will continue to grow, along with the waste they generate. People with the wealth to see the world will want to do so. The solution here is to apply circular economy concepts to production and consumption. Every material product or service that we consume should be produced with the least possible environmental impact, and after a product or service is consumed, the waste from that process should be used as an input to another productive process. With human ingenuity, engineering, and reasonable rules of the road (i.e., regulation), we can grow our economy while reducing its environmental impact. We have done this in the United States and much of Europe for about four decades. We do not need to do without to live sustainably. We can build our production and consumption around the principles of environmental sustainability.

We are going to need to do this because people in the developed world like the stuff they have, and people in the developing world want to enjoy the same stuff. The fact that more and more consumption is nonmaterial is an indication that our own needs and desires are not necessarily in conflict with environmental sustainability. But a failure to maintain and improve economic life around the world is a prescription for political instability. That instability could lead to violence and terror that can endanger both people and the planet. Economic issues can lead to instability and violence, but as we learned with Mr. Putin's invasion of Ukraine, political factors can lead to instability as well. One way that political strife affected environmental sustainability involved the Russian attacks on Ukraine's nuclear power plants and the land surrounding the site of the Chernobyl nuclear disaster. The technologies that we select to avoid greenhouse gases must be selected with care.

FOCUSING ON ENVIRONMENTALLY SUSTAINABLE ECONOMIC GROWTH

The strategy for reducing pollution and growing a renewable resource–based economy must focus on economic growth that does not harm the environment. Culture, entertainment, education, designing software, physical fitness, and wellness are all activities with low environmental impact. Manufacturing, construction, transportation, agriculture, and operating the built environment (heating, cooling, and lighting our homes, transport systems, and businesses) are higher-impact areas of economic activity that require creativity and ingenious engineering in order to reduce their impact. It can be done.

The agriculture sector is a good example of a part of the economy using cutting-edge technology to reduce both pollution and costs. Writing in October 2020, *Fortune* reporter Aaron Pressman observed:

> Some farmers have turned to a high-tech solution for improving yields and reducing costs. For example, harvesting combines introduced this year by Deere & Co. include high-resolution cameras and sensors linked to A.I. software. The system monitors grain as it's collected and adjusts dozens of settings on the combine in real-time to maximize how much grain is chopped from each stalk and to minimize waste. An X9, as the new machine is called, can harvest a field 45 percent faster than Deere's older equipment lacking the automated system and it uses 20 percent less fuel. Widespread use of precision agriculture methods could reduce farming costs by $100 billion while saving 180 billion cubic meters of water by 2030, according to a McKinsey study done for the World Economic Forum. Using data crunching and A.I. analysis could increase farmer incomes by another $70 billion by 2030, the study found. (Pressman 2020)

We see similar efforts in the fashion industry and many other types of manufacturing as companies seek to reduce both their environmental footprint and the cost of production. In agriculture, any fertilizer or pesticide that is not absorbed by plants and ends up in groundwater amounts to waste. If farmers can calibrate the use of these expensive substances more precisely, they can save money while protecting the planet. Manufacturing recycled materials can be used to make new clothing, and new technologies are being used to reduce the environmental impact of dyeing and manufacturing fabric. The key is that manufacturers and engineers add environmental impact to their design parameters when creating a machine or a manufacturing process. This enables both economic growth and environmental protection, and thousands of examples of this are now underway.

There are many examples of production processes and business practices that improve quality, reduce costs, and reduce environmental impact. The important part is designing processes that reduce environmental impact from the start. As I've noted elsewhere, one of the reasons that environmental protection is sometimes seen as adding cost is that the early environmental protection technologies were add-ons. If a process or facility is originally designed to reduce impact, costs tend to be lower.

CONNECT TO EQUITY AND
ENVIRONMENTAL JUSTICE GOALS

The strategy for growing a green economy must also be connected to equity and environmental justice goals. The jobs and clean environment created by the green economy cannot be limited to the privileged in our society. Many of the pieces of U.S. federal legislation proposed in 2021 to accelerate the transition to a green economy specified that a large proportion of subsides should be allocated for the communities most damaged by environmental impacts. These tend to be communities of color and low-income neighborhoods. The transition to a green economy will create many jobs—especially in the construction of infrastructure—for energy, waste, water, climate adaptation, and transportation. We must also ensure that the negative environmental impacts that these communities experience are prevented or remedied.

There is also an issue related to cross-generational equity. Climate change is a different issue for young people than for older people. When I first learned about global warming, I was reading projections during the

1980s and 1990s of impacts that would take place decades in the future. But that future is here now, and it is being experienced by all of us. The difference between older and younger people on climate change is that young people have never known a world without global warming.

The intense effects of extreme weather, rising average global temperatures, melting glaciers, and rising seas are no hoax, and young people know that the reality they see will get worse if action is not taken. The ethical imperative to act is clear. The issue is not whether we should act, but what actions we can take. What actions will protect the planet while maintaining the quality of life that we enjoy in the developed world and ensuring that everyone on the planet can enjoy those benefits as well? Those of us who have created this world and reaped its benefits are imposing the costs of those benefits on our children and grandchildren.

Today, we have increased income inequality, homelessness, and a struggling middle class. Many Americans have a pervasive feeling of insecurity about the future. Factory workers fear automation, students fear debt, and young people worry about how they will find meaningful work and survive on a warming planet. Many of us who will not live to see that climate future share our children's worries about it. Even if the ethical issues are not as crisp and clear as some climate activists like to think they are, they still have power and currency.

Ironically, a great deal of this uncertainty could be addressed by the mobilization required for a great national project. That work—a moonshot for the twenty-first century—could be the decarbonization of the energy system and the transition to a renewable resource–based economy in the United States. That great national project would provide a sense of purpose, along with employment. It could address equity and environmental sustainability simultaneously.

MOVING BEYOND SYMBOLS, GOALS, AND TARGETS TO A FOCUS ON IMPROVING CURRENT LEVELS OF PERFORMANCE

Climate change has generated scores of greenhouse emission targets in many nations, states, cities, and private organizations. There is literally an industry in aspirational (and largely unattainable) goals. There is a field of symbolic sustainability, in which an organization or a place aspires to present an image of sustainability. To such people and groups, sustainability is just a green branding exercise. This variant of sustainability makes no one's planet less polluted and no one's world more socially just, but it induces consumers to favor one product over another because of its green image. This "greenwashing" is far from a new story, but more and more companies have come to understand that people under thirty years of age are basing consumer decisions on the image that an organization presents as it sells its goods and services. Google once had the motto "Don't be evil," which was eventually replaced by "Do the right thing" when the company morphed into Alphabet. But the motto was tossed back at the company in the wake of its efforts to commercialize data about individual users. Companies like Google, Microsoft, and Amazon are pushing hard to reduce their carbon footprint and environmental impact—and they are making sure that we know it. They are investing in solar panels and taking other steps to reduce their environmental impact. But is it substantive, or simply symbolic? It is easy to be mistrustful of these efforts because we have no way of knowing what is really going on.

In a world dominated by media images and symbolism, the difficult work of sustainably managing our organizations can take a back seat to the easier job of looking like we are accomplishing something. The operational reality of organizational behavior requires that we really understand the tasks being performed. Who does what, when, and why? How does the organization obtain the resources needed to survive and thrive? My focus here is on managing what I call the "physical dimensions" of environmental sustainability: the use of energy; the source of energy, water, and materials; the reduction of waste and the environmental impact of an organization's production; and the impact of the consumption of whatever the organization produces. The key question to be addressed is: How does an organization learn to reduce its negative impact on the environment?

Sustainability management is the future of management and is the next phase of management innovation. Sustainability needs to be integrated into the heart of what organizations do. In the twentieth century, organizations constantly evolved, adding international operations, supply chains, information technology, sophisticated marketing, strategy, finance, accounting, and new forms of group processes. Today, organizations throughout the globe are attempting to reduce their carbon footprint, minimize climate and environmental risk, enhance resiliency, and incorporate sustainability into their operations. The problem is that in many cases, they do not know how to do it. Companies, governments, and nonprofits are recognizing that they need to take these actions, but their expertise lies in their own core functions, not in developing sustainability plans and assessing operational risk in this context. To have meaningful effects on environmental sustainability, corporations, nonprofits, and governments must bring it into the regular fabric of organizational life. In the long run, an organization should not need a "sustainability office"; every office should become a sustainability office. Managing the physical dimensions of sustainability must be added to standard operating procedures so all managers become sustainability managers. Big ideas, ambitious goals, and beautiful policies won't translate into on-the-ground change until organizations change.

The bigger problem that we are seeking to address ultimately comes down to the challenge of global sustainability: How can we create a high-throughput economy without destroying the planet? We are, quite literally, talking about saving the planet for our children. The issues at the heart of the sustainability problem are complex, global, and systemic, and they cannot be easily solved. These challenges require action by stakeholders at all levels—federal and local governments, multinational corporations, small businesses, quasi-government entities, and NGOs and nonprofits. But translating these ideas into action requires far greater organizational capacity than we currently have. We may know how to build a Leadership in Energy and Environmental Design (LEED)–certified, energy-efficient building, but do we know how to maintain it? Often, we don't. If we learn how to maintain the building, how do we make its continued performance a central organizational value?

This organizational challenge is not widely acknowledged; it is not the flashiest problem that environmental advocates are calling attention to. However, organizations are beginning to understand that they cannot simply create a sustainability office and expect the problem to be solved. Many of our best leaders know that they need to manage the physical dimensions of sustainability. There is ever-broader acceptance of *why* environmental sustainability is needed, but there is not yet a deep understanding of *how* to do it. What we need to do is learn how to build that capacity. Since I'm an academic, you will not be surprised to learn that I believe that the knowledge we need must come from academically rigorous, applied research.

As a student of management, I know that each organization and organizational environment is unique, but there are patterns and practices that can be identified and understood. Under what conditions do organizations tend to do better at reducing their use of energy and changing their source of energy from high carbon to low? When do investments in closed-systems engineering get made, and when are they shunned? What type of managers take environmental factors seriously, and what type think that it's a joke? In organizations heavily dependent on an extended supply chain, what strategies have succeeded when trying to influence the behavior of suppliers, and which have failed?

There has been a tendency for some environmental advocates to focus their attention on symbolic political battles, in part as a way of gaining attention and educating the public. The attraction to symbolism dates to the start of the modern environmental movement. According to Sam Whiting of *SF Gate*:

> Before there was Earth Day, there was the Survival Faire, an ecology extravaganza that did not survive for even one year. But its signature action is still remembered by those who were at San Jose State College in early 1970. That was when a bunch of students bought a brand-new, never-started Ford Maverick, then pushed it from the dealer's lot into the center of campus, where it was ceremoniously buried in a pit, 12 feet deep. The elaborate funeral procession was conceived of as an anti-internal combustion, anti-smog statement to cap off the weeklong fair, which itself resulted in one of the first environmental studies departments to be established at any university in the nation. (Whiting 2010)

Today's far more sophisticated versions of car burying include divesting pension funds from fossil fuel companies and suing energy companies for the environmental damage that they've caused. It worked with the tobacco companies, so why not the oil companies? It might work in some sense, and it certainly plays a role in public education. But the real work is convincing people to stop using tobacco and to use renewable sources of energy. Walking on the streets of New York, or cities anywhere in the world, I see plenty of people smoking. The tobacco companies seem to be doing all right. Driving down those same streets are plenty of vehicles powered by fossil fuels. We're all addicted to fossil fuels, and fossil fuel-based energy is central to our daily lives. I agree that investing in fossil fuel companies is a bad idea, but I still fill my tank with gasoline. Those companies get our money one way or the other. The real work is getting electric vehicle infrastructure built, advancing battery and renewable energy technology, and implementing these new technologies once they arrive. I am interested in learning how to encourage and accelerate *real* change. Symbolic gestures have a role to play, but we should not confuse them with operational reality. Creating an operational reality can be tedious work. It involves deep analysis of work processes to understand how they are performed and their environmental impacts. It then involves experimentation to reduce these impacts while maintaining production.

Another element of environmental symbolism is the setting of greenhouse gas reduction and other environmental targets: this much by 2030, that much by 2050. My university has set targets, so has my city, my state, many corporations, the United Nations—it's happening everywhere. Sadly, it means nothing. Aspirational goals have value, but I am far more interested in rigorously measuring our current level of performance and focusing operational attention and measurement on improvements. What specific steps can we take, now, to reduce emissions? Not what we hope to do next year, but what we can actually do this year.

We need to move away from an emphasis on symbols to operational reality. That is why metrics are important. Even if we can't develop universal environmental sustainability metrics, we can insist on independent and rigorous audits of measures. Once in place, government policy can reinforce improvements by providing tax credits or deductions as rewards for specific percentages of carefully measured and documented improvements. The SEC's proposed climate disclosure rules in 2022 may well be the opening step in the development of Generally Accepted Environmental Sustainability Metrics. If they manage to get past the proposal stage and companies must adhere to these rules, then environmental sustainability may well move beyond symbols and get closer to reality.

CONCLUSION: A STRATEGY FOR ENVIRONMENTAL SUSTAINABILITY

The strategy suggested in this chapter has four interconnected elements:

1. *Connect economic growth to environmental sustainability*: The important thing is to change the nature of consumption and to use technology to reduce environmental impacts. People do not need to sacrifice and receive fewer benefits—but consumption becomes a means rather than an end.

2. *Environmental sustainability is a key element of economic modernization*: We invest in new technology and infrastructure that makes our energy, water, waste, and transport systems more efficient, cleaner, and cheaper. We preserve land and ensure that its exploitation is environmentally sustainable and permits regeneration.

3. *While we must have environmental standards and rules, the emphasis is on achieving improvement through incentives rather than punishments*: This means no carbon taxes, but tax benefits for reducing pollution. Environmental criminals must be punished, but let's not build a system that assumes that everyone is a criminal.

4. *Focus on environmental and nonenvironmental sustainability metrics, build a sustainability management culture, and replace aspirational targets with rigorous measurement of current performance and an operational emphasis on continuous improvement.*

This strategy is built on a positive vision of environmental sustainability. Addressing issues like climate change and maintaining biodiversity does not require sacrifice but can instead result in more exciting and interesting sustainable lifestyles. Technological advances have been changing human lifestyles and enhancing our quality of life for centuries. Our task now is to steer those changes to enrich our lives while preserving the planet. The strategy is also built on a practical and realistic political analysis. Yes, we need to change the type of energy that we use, but making fossil-fuel energy more expensive will not drive people toward renewables unless they are as convenient and available as current energy sources. More expensive energy will generate political opposition to efforts to reduce greenhouse gases. In the long run, renewables will be cheaper than fossil fuels, so the best political strategy is to subsidize renewables until subsidies are no longer needed.

WORKS CITED

Appelbaum, Binnyamin, and Jim Tankersley. 2018. "The Trump Effect: Business, Antici-pating Less Regulation, Loosens Purse Strings." *New York Times*, January 1, 2018. https://www.nytimes.com/2018/01/01/us/politics/trump-businesses-regulation-economic-growth.html.

Ballard, Ed. 2022. "Disclosures on Nature, Climate Go Hand in Hand, Nestlé Risk Chief Says." *Wall Street Journal*, March 24, 2022. https://www.wsj.com/articles/disclosures-on-nature-climate-go-hand-in-hand-nestle-risk-chief-says-11648116000?mod=hp_minor_pos1.

Blunt, Katherine. 2022. "Utilities Plan Huge Electric Grid Upgrades . . ." *Wall Street Journal*, April 17, 2022. https://www.wsj.com/articles/utilities-plan-huge-electric-grid-upgrades-adding-to-power-bills-11650187802?mod=Searchresults_pos12&page=1.

Cho, Renee. 2017. "What the U.S. Military Is Doing about Climate Change." *State of the Planet*, September 20, 2017. https://news.climate.columbia.edu/2017/09/20/what-the-u-s-military-is-doing-about-climate-change/.

Cohen, Steven, Sheldon Kamieniecki, and Matthew Cahn. 2005. *Strategic Planning in Environmental Regulation: A Policy Approach That Works*. Cambridge, MA: MIT Press.

Dell, Rebecca. 2021. "Making the Concrete and Steel We Need Doesn't Have to Bake the Planet." *New York Times*, March 4, 2021. https://www.nytimes.com/2021/03/04/opinion/climate-change-infrastructure.html.

Environmental Protection Agency (EPA). n.d. "Air Quality—National Summary." Accessed March 31, 2022, from https://www.epa.gov/air-trends/air-quality-national-summary.

European Commission, Directorate General for Research and Innovation. 2021. *Incentives to Boost the Circular Economy: A Guide for Public Authorities*. Luxembourg: Publications Office of the European Union. https://data.europa.eu/doi/10.2777/794570.

Green, Jessica F. 2021. "Does Carbon Pricing Reduce Emissions? A Review of Ex-Post Analyses." *Environmental Research Letters* 16 (4): 043004. https://doi.org/10.1088/1748-9326/abdae9.

Ioannou, Lori, and Magdalena Petrova. 2019. "America Is Drowning in Garbage. Now Robots Are Being Put on Duty to Help Solve the Recycling Crisis." *CNBC*, July 27, 2019. https://www.cnbc.com/2019/07/26/meet-the-robots-being-used-to-help-solve-americas-recycling-crisis.html.

Lindstrom, Anodyne, and Sara Hoff. 2021. "U.S. Electricity Customers Experienced Eight Hours of Power Interruptions in 2020." U.S. Energy Information Agency, November 10, 2021. https://www.eia.gov/todayinenergy/detail.php?id=50316.

Milbrath, Lester W. 1989. *Envisioning a Sustainable Society: Learning Our Way Out.* SUNY Series in Environmental Public Policy. Albany: State University of New York Press.

Osborne, David, and Ted Gaebler. 1993. *Reinventing Government: How the Entrepreneurial Spirit is Transforming the Public Sector.* New York: Penguin.

Patten, Rachel, Hayley Herzog, Griffin Smith, Andrew Solfest, Anna Beeman, Jenny Bock, et al. 2021. "HR 2759: The Department of Defense Climate Resiliency and Readiness Act." Columbia University, School of International and Public Affairs. https://00b4e11a-a0e5-44f0-a4c5-d4dfe41c1ebc.filesusr.com/ugd/571f98_9975b7735 0f047a18e3d591ba7e6c392.pdf.

Pressman, Aaron. 2020. "A.I. Gets Down in the Dirt as Precision Agriculture Takes Off." *Fortune*, October 5, 2020. https://fortune.com/2020/10/05/a-i-precision-agriculture -deere/.

Schwarzenegger, Arnold. 2022. "Schwarzenegger: We Put Solar Panels on 1 Million Roofs in California. That Win Is Now Under Threat." *New York Times*, January 17, 2022. https://www.nytimes.com/2022/01/17/opinion/schwarzenegger-solar-power -california.html.

Securities and Exchange Commission. 2022. "SEC Proposes Rules to Enhance and Standardize Climate-Related Disclosures for Investors." https://www.sec.gov/news /press-release/2022-46.

Sustainability Accounting Standards Board (SASB). 2022. "SASB Standards & Other ESG Frameworks." Value Reporting Foundation. https://www.sasb.org/about/sasb -and-other-esg-frameworks/.

Tita, Bob. 2022. "Aluminum Makers Seek Old Cans, Shredded Cars to Fuel New Plants." *Wall Street Journal*, March 26, 2022. https://www.wsj.com/articles/aluminum-makers -seek-old-cans-shredded-cars-to-fuel-new-plants-11648287002?mod=hp_lead _pos2.

Vanderford, Richard. 2022. "SEC Climate Disclosure Proposal Looms as Litigation Risk." *Wall Street Journal*, March 26, 2022. https://www.wsj.com/articles/sec-climate -disclosure-proposal-looms-as-litigation-risk-11648299600.

Vincent, Carol Hardy, Laura Hanson, and Lucas Bermejo. 2020. "Federal Land Ownership: Overview and Data." Congressional Research Service, February 21, 2020. https://sgp.fas.org/crs/misc/R42346.pdf.

White House. 2021a. "FACT SHEET: The Bipartisan Infrastructure Investment and Jobs Act Advances President Biden's Climate Agenda," August 5, 2021. https://www .whitehouse.gov/briefing-room/statements-releases/2021/08/05/fact-sheet-the -bipartisan-infrastructure-investment-and-jobs-act-advances-president-bidens -climate-agenda/.

White House. 2021b. "FACT SHEET: President Biden Signs Executive Order Catalyzing America's Clean Energy Economy Through Federal Sustainability," December 8, 2021. https://www.whitehouse.gov/briefing-room/statements-releases/2021/12/08/fact-sheet-president-biden-signs-executive-order-catalyzing-americas-clean-energy-economy-through-federal-sustainability/.

Whiting, Sam. 2010. "San Jose Car Burial Put Ecological Era in Gear." *SF Gate*, April 20, 2010. https://www.sfgate.com/green/article/San-Jose-car-burial-put-ecological-era-in-gear-3266993.php.

Wilts, Henning, Beatriz Riesco Garcia, Rebeca Guerra Garlito, Laura Saralegui Gómez, and Elisabet González Prieto. 2021. "Artificial Intelligence in the Sorting of Municipal Waste as an Enabler of the Circular Economy." *Resources* 10 (4): 28. https://doi.org/10.3390/resources10040028.

World Economic Forum and Deloitte. 2020. "Statement of Intent to Work Together towards Comprehensive Corporate Reporting." Impact Management Project. https://29kjwb3armds2g3gi4lq2sx1-wpengine.netdna-ssl.com/wp-content/uploads/Statement-of-Intent-to-Work-Together-Towards-Comprehensive-Corporate-Reporting.pdf.

5

BUILDING PUBLIC-SECTOR INFRASTRUCTURE TO SUPPORT ENVIRONMENTAL SUSTAINABILITY

The transition of the American economy to one that is environmentally sustainable requires an approach that includes a positive political strategy, technological innovation and research, private sector entrepreneurship, regulation, government procurement, and finally government investment. This chapter focuses on investment in infrastructure. Where the infrastructure is capable of generating user fees or other revenues, then public investment is used to stimulate private investment. Where the infrastructure generates no revenue or insufficient revenue, then rather than seek a cross-sector collaboration, government will proceed on its own. In the United States, investment of authority and funds in the stewardship of the government's vast land resources must also be a key element of a strategy for environmental sustainability. In addition to these physical infrastructure investments, as an educator I need to stress the need for human capital management in the field of sustainability science, policy, and management.

For decades, the federal government has disinvested in infrastructure. Airports, bridges, roads, and rails are in disrepair. Our water and sewage systems are outdated. In some rural areas, septic tanks and fields are poorly managed, and in poor areas of the South, household waste is piped into backyards. Our energy system is vulnerable, wasteful, and in desperate need of modernization. Our waste management system still depends on landfilling when it could be modernized to facilitate resource recovery. On a planet with so many people, the resources that we need to provide food, clothing, and shelter cannot continue to depend on what we can extract from the planet. We need to mine our waste stream for materials that can be repurposed into food, clothing, and shelter. As raw materials become more expensive to find and mine, reprocessed materials will become economically competitive. To facilitate the circular economy, research and development, and then infrastructure investment, must be devoted to waste sorting and reprocessing.

Collective goods require collective resources, and the United States has been on an antitax binge since Ronald Reagan's administration. The good news about the state of disrepair is that when we finally develop the political will to rebuild, we can now invest in technology that will advance environmental sustainability.

The areas of investment discussed in this chapter include the following:

- Energy
- Water
- Waste
- Transportation
- Public health
- Communications

Our existing infrastructure is falling apart. Reconstruction of rails, ports, roads, tunnels, urban and national parks, schools, libraries, and the energy grid would provide far more economic and societal benefits, and with far lower economic, environmental, and human health costs. All infrastructure investment is not created equal. We could invest in bike paths and mass transit if we are looking for new projects to build. The best shovel-ready projects tend to be those that are rebuilding infrastructure that is falling apart. Spend money and put people to work getting rid of lead water pipes. Spend money and put people to work installing solar panels and building smart grids. Create public-private partnerships to design, build, and manage rebuilt facilities. Create a climate corps, and put young people to work preserving and managing our forests and government-controlled lands.

The problem is that many of the ideologues serving as elected officials do not seem to recognize the importance of public capital investment in infrastructure. FDR's New Deal during the Great Depression built dams, parks, airports, roads, and bridges and provided rural electrification. People were put to work building projects that provided continuous benefits to society. The Civilian Conservation Corps put hundreds of thousands of unemployed people to work building and managing parks and public lands in the United States. The idea was not to make money for corporations, although plenty of them made money, but to provide benefits to the public. That type of investment is needed to transform our economy into a renewable resource–based economy.

The infrastructure strategy should provide defensive investment to prevent old facilities from falling apart, but also allocate creative investment that makes our economy more efficient and competitive:

1. **Defensive investment** includes modernizing aging transportation, water, park, waste, and sewage facilities that are in danger of collapsing due to age or extreme weather, or could be made more cost-effective by modernizing with new technology.
2. **Creative investment** includes putting capital into the energy system: the grid, renewable energy generation, and storage and electric vehicle charging stations. It also includes finding new transportation facilities that increase the efficiency of moving people and goods. Finally, it would include investments in automatic waste sorting that would enable us to mine resources from our waste.

The focus should be on infrastructure that makes our economy more globally competitive. Renewable energy and smart grids reduce greenhouse gases, but they also make our energy supply less expensive and more reliable. Modern water, waste, and sewage technology reduces the cost of dealing with the biological necessities of human settlements. Old water systems leak, waste water, and can pollute drinking water (as happened in Flint, Michigan). Inadequate sewage treatment leads to expensive flooding and can lead to higher-cost water filtration.

In addition to these specific areas of infrastructure, we must protect our investment by building infrastructure so it can handle the impacts of climate-induced extreme weather events. This requires flood control, shoreline defenses, forest management, and buildings and other infrastructure that can withstand wind, flood, cold, and heat. I should also mention that the infrastructure needed for environmental sustainability is not the only place where we must focus our efforts. I do not discuss housing or education here, although clearly both areas require investment.

ENERGY

Our energy system is highly centralized and vulnerable to disruption by weather, online hackers, and deterioration due to old age. A modern smart grid, composed of thousands of connected but independent microgrids, can allow mass sharing of renewable energy and match energy generation to utilization. This system will be expensive to build but far less expensive to operate. It would also be more reliable than the current system, eliminating many energy blackouts. We are increasingly dependent on energy in our daily lives, and the promise of less expensive and more reliable electricity is a political winner.

There are three main areas of infrastructure investment needed for renewable energy:

1. Solar, wind, geothermal, and hydrogeneration of power
2. Smart grids, microgrids, and distributed generation of energy
3. High-voltage, long-distance distribution lines

While the cost of transforming the electric grid will be high, renewable energy is becoming less expensive and is already competitive with fossil fuels. Public infrastructure investment, along with solar cell cost reductions, should help transform the private energy market. Accelerating the decarbonization process so that it occurs faster than the market would on its own will require both government incentives and regulations.

Solar, wind, and geothermal power will benefit from government investment in basic research and tax incentives for the private sector to commercialize new technologies. Opportunities for hydrogeneration should be pursued, but with greater emphasis on the ecological costs of dam construction. Taking advantage of tides and river currents should also be explored, again with an understanding of the ecological impacts. Investments in solar and wind farms and water-based power plants should be subsidized and initiated by governments, particularly where there is a reasonable prospect that a long-term revenue stream will develop that will allow governments to withdraw their involvement.

Investments should both modernize and decentralize the gird, and enable homeowners to disconnect or reduce their dependence on the grid. It is obvious that individuals and families will eventually be able to generate and store enough energy for their daily use. Improvements in battery technology and solar cells should make this possible. At that point, people may disconnect from the grid or maintain it as a backup energy source. Institutions like hospitals, schools, and manufacturing facilities may prefer the reliability and greater capacity of the grid. In any case, we need to assume that a modernized, cost-effective electric grid will remain an essential public good for the foreseeable future. Investment in the grid and its modernization should be a high priority.

In American cities, energy can become more decentralized via neighborhood microgrids and distributed generation of energy. The reliability of our centralized energy system will decline due to age and extreme weather, and we need to build in redundancy and the ability to uncouple from the system when it goes down. Until we develop technology that will incorporate solar arrays into the windows of apartment buildings, we need to prevail upon single- and multiple-family homes to install rooftop solar cells and energy storage batteries. Parking lots, big box stores, shopping malls, and other large structures should be covered by solar arrays. State and local governments should incentivize solar installation through property tax reductions and no-interest loans. Large-scale adoption of solar power combined with wind power can help us reduce our dependence on fossil fuels. Offshore windmills must be constructed at a far faster pace than we have seen to date. We will need additional technology to do the rest of the job, but let's start with what we have. According to the U.S. Energy Information Administration:

> In 2022, we expect 46.1 gigawatts (GW) of new utility-scale electric generating capacity to be added to the U.S. power grid, according to our Preliminary Monthly Electric Generator Inventory. Almost half of the planned 2022 capacity additions are solar, followed by natural gas at 21 percent and wind at 17 percent. . . . Solar power will account for nearly half of new U.S. electric generating capacity in 2022. (Fasching and Ray 2022)

In both government and private buildings, the cost side of the energy transition includes capital investment in meters, energy controls, energy storage, solar and wind generation, building and window insulation, and similar items. It also includes training and staffing costs, along with the cost of developing and implementing new management systems. All of these investments reduce not only pollution, but also energy costs. Fuel costs go down, along with societywide savings due to the health and productivity benefits that occur with reductions in air pollution.

A final area of infrastructure investment is high-voltage, long-distance distribution lines that can carry electricity efficiently from rural solar and wind farms to population centers. This investment assumes that we will maintain a modern electric grid that will be powered by renewable energy. Oceans and deserts make excellent locations for wind and solar farms, but then that energy must be transferred to where it is needed. Current electric lines lose too much energy during transmission, but new high-voltage lines lose far less. A public-private partnership could get this process started, replaced eventually by private investment financed by user fees.

While our focus should initially be on the United States, at some point we need to take a more global approach. The only way to decarbonize the developing world's economy is to help developing nations leapfrog the fossil fuel phase of economic development by subsidizing renewables. China may be building coal-fired power plants, so we will need to subsidize their development of windmills, solar cells, and batteries. My hope is that then China will compete with us on renewables and leave the coal in the ground where it belongs.

There are many policy analysts and scientists who believe that nuclear power is the only way to decarbonize our energy system. I do not. As Barry Commoner (1979) once wrote: "Nuclear power—is . . . certainly an exotic way to boil water." Any technology that can damage ecosystems for thousands of years is to be avoided if at all possible. While the likelihood of an accident is low, the impact of an accident is unacceptably high. Any technology managed by humans—in other words, all technology— should assume that at some point, Murphy's Law will come into play: If something bad can happen, it will.

I understand that nuclear plants can generate power without increasing greenhouse gases, but the toxicity of nuclear fuel and waste makes nuclear technology too risky to rely on. Nuclear power is very capital intensive, and regulatory requirements for multiple safety systems increase the costs. The need for coolant means that nuclear plants are located near water, and the growing competition for waterfront sites for residential and commercial development is driving up the price of that land. As mentioned earlier, the origin of nuclear power, the "Atoms for Peace" program starting in the early 1950s, was an effort to change the image of nuclear technology. But it's always looked better in theory than in reality.

WATER AND SEWAGE TREATMENT

There was a time when our population was small, and groundwater was so clean that many people could simply dig a well on their property and draw clean and plentiful water from that source. According to the United States Geological Survey (USGS):

> More than 43 million people—about 15 percent of the U.S. population— rely on domestic (private) wells as their source of drinking water. The quality and safety of water from domestic wells are not regulated by the Federal Safe Drinking Water Act or, in most cases, by state laws. Instead, individual homeowners are responsible for maintaining their domestic well systems and for monitoring water quality. (USGS 2019)

The USGS estimated that about 20 percent of these wells did not meet federal water quality standards. As our groundwater supply becomes contaminated by pollutants from a variety of sources, and as supplies are affected by climate-induced changes in weather, the number of people depending on private water sources will drop.

In New York City, the pollution of groundwater took place well over a century ago. Since the nearby water sources were largely salinated, the city had to look north for uncontaminated fresh water and began building a system of reservoirs, water tunnels, and pipes. The city's first reservoir was at the site of the current Bryant Park and New York Public Library. Its next reservoir was located in Central Park, but as these early sites became contaminated, land north of the city was purchased and developed as reservoirs and water tunnels. A massive water system was then created, without which the city of New York would never have been able to develop and grow as it did.

Today, 85 percent of the American people depend on public water systems. Much of the water supply requires extensive filtration. In addition, water can be contaminated when it is piped from a water plant to or within a building if the pipes are old. The construction and updating of water filtration plants require large amounts of capital. The federal government has continued modest investment in the form of low-interest loans in water supply and sewage treatment, even during Donald Trump's administration. According to Jonathan L. Ramseur of the Congressional Research Service:

> The Clean Water Act (CWA) establishes performance levels to be attained by municipal sewage treatment plants in order to prevent the discharge of harmful wastes into surface waters. The act also provides financial assistance so that communities can construct treatment facilities and related equipment to comply with the law. Although approximately $104 billion in CWA assistance has been provided since 1972, funding needs for wastewater infrastructure remain high. The Environmental Protection Agency (EPA) estimates that the nation's wastewater treatment facilities will need $271 billion over the next 20 years to meet the CWA's water quality objectives. (Ramseur 2018)

Federal support for water systems and sewage treatment was substantial in the early days of the water pollution control program in the 1970s, but it eventually ended up as state and local responsibility. Modern development patterns have added significantly to the need for water infrastructure and a federal role has continued, however inadequate it may be. Again, according to Ramseur:

> In both FY2016 and FY2017, Congress provided $1.394 billion for the Clean Water State Revolving Fund (CWSRF) program. However, funding for the program increased by 22 percent in FY2018. The Consolidated Appropriations Act, 2018 (P.L. 115–141) provided $1.694 billion to the CWSRF program. In addition, Congress established the Water Infrastructure Finance and Innovation Act (WIFIA) program in 2014 (P.L. 113–121).

WIFIA provides direct loans for an array of water infrastructure projects, including CWSRF-eligible projects. EPA issued its first WIFIA loan in April 2018. In FY2018, Congress appropriated $63 million to EPA for the WIFIA program (roughly double the FY2017 appropriation). EPA estimates that this funding will provide approximately $5.5 billion in credit assistance. (Ramseur 2018)

These funds do not come close to meeting national needs. President Joe Biden's bipartisan infrastructure bill in 2021 provided considerably more funding in the form of grants for state and local water infrastructure. But the need for additional funding is great and will be exacerbated by the impact of climate change on water systems.

For example, in the western part of the country, the massive Colorado River Project provides water from the Rocky Mountains to dry climates in the Southwest. That infrastructure is under strain due to drought and changing patterns of rainfall. According to climate reporter Lauren Sommer, writing for NPR:

Historical climate data such as river flows and rainfall totals told engineers how big to build reservoirs and canals. The data also told them how much water was available to divide up among cities and farms. Climate change is putting that system under increasing stress, shrinking water supplies for tens of millions of people and for the farmland that produces most of the country's fruits and vegetables. Water cutbacks are reverberating through California's $50 billion agricultural industry, which employs tens of thousands of people in many small towns. (Sommer 2021)

The need for desalination and recycled wastewater will grow in the coming years, and we can expect that funding will be required to meet the capital costs of the infrastructure to address these issues. Low-cost and even free water supplies will no longer be common as massive investment in water systems becomes required. Just as New York City faced this need over a century ago, and the Water Authority charges New York property owners steadily increasing user fees, we can expect a similar pattern to emerge throughout the United States. The days of free water are coming to an end.

WASTE

According to *Washington Post* reporter Tik Root, citing a study from the National Academy of Sciences, the United States is the world's worst plastic polluter:

> The United States contributes more to this deluge than any other nation, according to the analysis, generating about 287 pounds of plastics per person. Overall, the United States produced 42 million metric tons of plastic waste in 2016—almost twice as much as China, and more than the entire European Union combined. . . . The vast majority of plastics are made from fossil fuels, and some can take hundreds of years to decompose. The researchers estimated that between 1.13 million to 2.24 million metric tons of the United States' plastic waste leak into the environment each year. About 8 million metric tons of plastic end up in the ocean a year, and under the current trajectory that number could climb to 53 million by the end of the decade. (Root 2021)

Solid waste management is a challenge for large urban areas around the world. Removing garbage from residential, institutional, and commercial locations in cities is a major logistical and operational task. Waste management is usually a function of local government and a major item in a city's budget. Solid waste generation rates are rising fast, particularly in cities experiencing increasing population rates and higher economic activity, putting pressure on municipal governments to deal with rising costs and environmental impacts.

The waste from cities around the world is rapidly increasing. According to the World Bank: "In 1900, the world had 220 million urban residents that produced 300,000 tons of waste per day; by 2000, those numbers grew to 2.9 billion people generating 3 million tons of solid waste per day. Worldwide, waste rates are expected to triple by 2100, exceeding 11 million tons per day" (Goto 2013).

The global cost of processing trash is rising too, from $205 billion a year in 2010 to $375 billion by 2025. A majority of the increased costs is being incurred in the developing world.

The global increase in waste material is resulting in an increasing amount of waste that is recycled, burned for energy, or, in the case of food waste, reprocessed as fertilizer. East Asia is now the world's fastest-growing region for creating waste. Waste generation in Asia's urban areas is expected to soon reach 1.8 million tons per day. In 2004, China surpassed the United States as the world's largest waste generator. The Chinese government has developed a number of laws and plans related to waste management. China's twelfth Five-Year Plan sets ambitious goals for managing solid waste, with an emphasis on recycling. However, China is experiencing rapid growth in waste generation. According to the World Bank, the quantity of municipal solid waste generated in China's cities increased more than fivefold between 1980 and 2009, from 85,000 tons to 430,000 tons per day, and is projected to reach 1.6 million tons per day by 2030.

Most waste in China goes into landfills or unregulated waste heaps outside major cities, and as China's landfills are filling up, cities are starting to burn waste to generate electricity at waste-to-energy plants. Overall, in China, waste management is rapidly shifting from the use of landfills to incineration. In 2017, 84.6 million metric tons of waste were burned and 120.3 million metric tons were landfilled. By 2019, 121.7 metric tons were incinerated compared to 109.5 million metric tons being landfilled.

However, there is increasing public concern about the environmental performance of these waste incinerators and their impact on the local environment and communities. While many waste-to-energy plants can burn garbage with little pollution, cheap incinerators without pollution controls create massive amounts of air emissions.

In the United States, our per capita generation of waste peaked at the turn of the twenty-first century, but our total amount of waste continues to grow along with our population. In the Midwest and West, where land is relatively plentiful, most garbage is dumped into landfills. But a growing percentage of our garbage is recycled or burned in low-polluting waste-to-energy plants. In New York, the most crowded U.S. city (and possibly a taste of the future for the rest of the country), waste management has been a growing problem.

New York City's over eight million residents and millions of businesses, construction projects, and nonresident employees generate over fourteen million tons of waste and recyclables per year. This amount is so vast that waste is handled by two separate systems—one public and one private. The public agency, the New York City Department of Sanitation

(DSNY), serves residential buildings, government agencies, and many nonprofit organizations. Private commercial firms do not receive free garbage pickup by the city government. They must pay private companies to remove their solid waste. Public spending on garbage pickup and disposal is over $2.5 billion of the city's $100 billion annual budget. In 2000, when New York City still had its own landfill, the cost was $658 million. These costs do not include the money paid by commercial firms to private waste carters.

As with many other sustainability issues, one element of the problem is people's values and behavior. People like to use stuff, and when they are finished with that stuff, they need to throw it out. There are limits to the amount of time and energy that most people are willing to devote to managing their own garbage. There are also system limits to what is possible. If you separate your home garbage but the city has no real recycling program, your efforts have been wasted. If you live in the countryside, you may be able to compost your food waste, but in a city, you must depend on a special collection system and an anaerobic digester (a technology that mimics a compost heap) to convert your food waste to fertilizer.

One solution to waste management is nontechnological. It involves designing products that can be easily reconditioned and reused, and a postconsumption process that brings the product back to the manufacturer. Xerox does this by leasing some of its copiers and designing them for remanufacturing. Hewlett-Packard (HP) does it by designing its toner cartridges to be easily collected and then refilled. Just as economic development creates a demand for more energy and exacerbates the climate crisis, increased consumption results in more waste. With growing wealth, we will see growing garbage.

Unlike with climate change, which attracts billionaires like Bill Gates and celebrities too numerous to name, no one wants to do a benefit concert for garbage. No one wants a waste management facility near their home, and just about everyone hopes those big plastic trash bags can be magically transported to solid waste heaven. The solution to the climate change issue will be new renewable energy technology that drives fossil fuels from the marketplace.

Similarly, the solution to waste management will rely on new technologies. One of the most promising of these allows the collection of a single waste stream and then mechanically separates the garbage. With artificial intelligence (AI) and automation, we can expect this infrastructure to become operational and cost effective over the next decade. Today some of the sorted waste goes to an anaerobic digester, some is recycled, some is burned for energy, and the residue of the incinerated garbage can be used as construction material. However, we can expect advances in sorting infrastructure.

The specific infrastructure advances needed would include waste-sorting plants to separate food, plastics, paper, metals, and chemicals, and then sending these clean waste streams to reprocessing plants. These plants will be the logical connection points to a true circular economy. While some American cities have very low waste management costs, all of them pay to collect and dispose of waste. Those costs can be shifted to a system that, instead of dumping and burning waste, will sort and sell raw materials such as plastic feedstock, paper, and the chemical components of fertilizer. These reclaimed raw materials could pay for some of the costs of disposal in the future. But the cost of this sophisticated system of waste management will be large, and in its initial stages, it will require federal subsidies for research, pilot projects, and the cost of capital.

In addition to solid waste, we produce another waste product: sewage. There are three environmental sustainability goals that modern sewage treatment plants must evolve to reach. First, the energy that they use should be renewable. Second, the water that they return to waterways should be free of contaminants. (That is why sewage treatment is considered both waste and water infrastructure.) Third, the nutrients and other useful chemicals in the waste stream should be processed and purified to enable reuse as fertilizer. Our investment in sewage treatment in many parts of the United States is less than half a century old.

The technology of sewage treatment has advanced steadily, but many local governments do not have the funding or political will to do any more than what is absolutely required under law. In some rural areas, raw household sewage is piped onto land, and household septic systems and fields are poorly managed and regulated. Like garbage, investment in infrastructure is needed, but elected officials do not want to be linked to the issue. It will probably take a Love Canal–style environmental catastrophe to get rural sewage (or even urban sewage) systems onto the political agenda. As discussed in chapter 4, advances in waste sorting and mining, powered by automation and AI, are a central part of the strategy of moving to an environmentally sustainable economy. This strategy will require massive infrastructure investment.

TRANSPORTATION

Federal infrastructure plans imbedded in the 2021 trillion-dollar infrastructure bill mandated traditional road and bridge construction, along with improvements to intercity train and urban-suburban mass transit. Given the suburban patterns of land development in the United States, the central environmental sustainability issue is personal transportation rather than mass transportation. The electric vehicle is central to that transformation, and with some small additional subsidies in the beginning, market forces will result in the retirement of the internal combustion engine.

Mass Transit

While we need to invest in sustainable personal transit, some cities have the density to support mass transit. Cities require effective and efficient transportation systems because their fundamental purpose is to bring people together for economic, social, educational, and cultural purposes. Even though modern communication technology allows many of us to work from home all the time, humans crave social interaction, and many of our work and creative processes require face-to-face contact. COVID-19 increased the number of people working from home, but many office workers prefer a hybrid arrangement with some days in the office and some days at home. If city neighborhoods become isolated due to congestion—something that we see in places like Los Angeles—that interaction becomes costlier. People need to be able to move around the city quickly, inexpensively, and comfortably. New York City's subway system enables that to happen fairly quickly and cheaply, if perhaps without the comfort, but without the subway and bus systems, the city's economy and attractiveness as a place to live would collapse. New York's path toward long-term sustainability would abruptly end.

To keep the price of mass transit reasonable, it must be subsidized, and in recent decades, that subsidy has not kept pace with the needs. Congestion pricing is an elegant answer to that difficult question. In the next several years, motor vehicles in New York City's Central Business District are scheduled to begin paying a fee that will be allocated to mass transit. In New York, the city's subway system is like a human circulatory system, delivering blood to key organs. Its deterioration threatens the city's economic, cultural, and social life. Its revival, along with a modernized water, energy, and waste system, will help ensure the long-term environmental sustainability of the city.

The Electric Vehicle

The key ingredients for the decarbonization of the American economy are renewable energy and the electric car. Most of the use of fossil fuels in the United States takes place in transportation, and about 28.5 percent of greenhouse gases come from there. Close behind is electric power generation, with 28.4 percent of emissions. But as our electric system becomes more renewable, the proportion of greenhouse gases coming from transportation will increase if we don't move toward electric vehicles. Transportation's dominance of greenhouse gas pollution is a particularly American problem.

Electric vehicles cost less to run and maintain than cars with internal combustion engines. They have fewer moving parts and lower repair costs, and their fuel is less expensive. Electric vehicle manufacturers are targeting young people who have grown up in an ever-changing world based on advanced information technology. These cars are loaded with driver-friendly electronics such as a global positioning system (GPS), satellite radio, and cell phone technology.

For many Americans living in suburban homes with driveways and garages, the ability to charge their electric vehicles already exists. They simply need to clear some of the junk that they are storing in their garage and find an outlet to plug into. For those of us living in cities or people traveling away from home, public charging stations must be constructed. Parking lots and garages are the logical places to install these stations, and credit cards or smartphones could be used to enable drivers to pay for electricity. In some places, on-street parking meters might be converted to charging stations. But most of these facilities must still be built. Tesla has already created its own "supercharger network," with 2,000 locations and 20,000 chargers. It claims that the cost of a charge is less than the cost of a gasoline fill-up. While Tesla's charging network may work for a niche market, it is far from sufficient for the mass market. To accelerate the mass market's adoption of this technology, some of the infrastructure will require public subsidies, or at least a public capital investment to be repaid by vehicle charging fees.

In addition to charging infrastructure, the American consumer must be convinced that the electric vehicle is better than the internal combustion engine vehicle. For a generation, both will exist side by side, and we should expect that early on, every failed electric vehicle will be a news story. Repair shops that make their living off of today's motor vehicles may not make as much money with electric vehicles since they have fewer moving parts and require less repair work. In addition, auto mechanics will need to learn the new technologies of these vehicles, and some will be unwilling to retrain. Auto repair shops have learned how to repair modern computer-laden cars, so there is reason to believe that many would adjust themselves further to electric vehicles. Auto companies know how to market vehicles, and the billions of dollars and euros on the line will motivate them to learn how to message the advantages of the new models that they are building. But the process will take time, and the only way to speed it up is for government to provide subsidies to consumers.

The electric vehicle will take at least a generation to displace the internal combustion vehicle. My next car will be electric, but my current one has only 20,000 miles on it. When I trade it in, someone will buy it and use it. That someone could be the government, which then would need to send it to the fossil fuel graveyard. If the government doesn't buy and retire my car, it will emit greenhouse gases for at least a decade after I sell it.

Sustainable Air and High-Speed Rail

China and Europe are far ahead of the United States in developing high-speed rail. Investment in the 2021 infrastructure bill may provide faster train service between Washington, D.C., and Boston, but efforts in California appear doomed, and the rest of the country seems more interested in air than train travel. Research into renewable jet fuel and electric air travel is underway, but it should be a target for federally funded research. High-speed rail makes sense in parts of the United States, but like personal transport and electric vehicles, we might be better if we focus on environmentally sustainable air travel due to Americans' general resistance to train travel. (I should note that I much prefer train to air travel, so long as the time lost is not significant. But clearly, I am out of the American mainstream.)

PUBLIC HEALTH

The COVID-19 pandemic has highlighted the need for a permanent local, national, and global system of public health. Our world economy and society are now interconnected by supply chains, communications technology, and travel. Barry Commoner taught us that everything in the biosphere was connected to everything else, but now everything in human society is connected to everything else. And so the oceans and vast landmasses of this planet no longer protect us from the negative impact of human behaviors and technologies. Perhaps at one time the oceans protected the New World, but the nuclear bomb and intercontinental ballistic missile ended that idea for all time. Today, a disease that may have jumped from animals to humans in a food market in China has led to illness, death, and mass change in every corner of the planet. That is what I mean when I say we are an interconnected world. Just as we see the same images and buy the same products, we share the same viruses.

Global supply chains are a fact of economic life. They are built on geographic, historic, and cultural distinctions between people and places. Different places and people come to specialize in producing different things. We are able to bring those specialties together in a supply chain. These chains rely on inexpensive communication, information, and transportation and enable higher-quality and lower-priced goods and services. The economic and technological forces behind these trends are irresistible. Nationalist political leaders will lose the battle to influence global corporations if their political strategies do not include an understanding of the economic benefits of globalization. Supply chains will evolve into supply webs as organizations seek to ensure supply redundancies, but the economic logic of organizational specialization and networks will persist.

COVID-19 should be seen as a symptom of a global economy that we all depend on and benefit from. In other words, we will see more of these instances in the future. While we do not fully understand this particular virus, we have methods for analyzing its causes and effects and for reducing its transmission. Conducting the research needed to protect public health requires resources, expertise, and institutional capacity. The United States has the resources, expertise, and ability to develop the capacity to contain this disease, but to do so, we need calm, determined, and unified national leadership.

I hope that by the time these words are published, we will have gained full control of this pandemic; however, we will need to do a much better job of containing the next one. Each new disease must be studied to learn how to prevent and treat it. That means that we must spend more on medical research—and the effort must be global and far better managed than the one that we've seen thus far with COVID-19. While new vaccines were developed rapidly, their distribution and use were poorly managed, as was the process of rapidly isolating those infected early in the pandemic. Second, when we identify a new virus, the traditional public health techniques of testing, tracing, and isolation must be implemented immediately and aggressively. Until we have a way to prevent or treat a new disease, this must become as routine as going through the security line at the airport. This requires organizational capacity that is well-resourced, competent, global, and apolitical. That will not be easy to do. But trillions of dollars and other global currencies have gone into the COVID-19 response and impact, and certainly preventing another pandemic is motivation enough.

The global public health system that we need will not be appealing. Just as post-9/11 security was intrusive and inconvenient, health security will probably be worse. And one of the costs of that security will be yet another piece of our liberty. This is a real trade-off. Just as our cities and buildings are loaded with security cameras, and just as we are questioned when we enter airplanes and some office buildings, we will soon find our bodies scanned not just for weapons, but for disease as well. I am not happy about any of this, but the lockdowns of COVID-19 were no picnic either.

One of the great benefits of the sustainable lifestyle is its emphasis on social engagement and interaction. Cities enable enjoyment of public spaces, the arts, and the diversity of a city's neighborhoods. The goal is to engage in experiences rather than ownership and consumption. This enables economic activity with relatively low levels of environmental impact. People enjoy parks instead of private gardens; theaters instead of private screening rooms; mass transit, walking, or biking instead of personal transport. They view rather than own original works of art and live in smaller private spaces due to their easy access to diverse public spaces. Social distancing makes it impossible to share resources and experiences (so much for the sharing economy).

One part of the global health system will be the subsidization of vaccination and testing by the nations able to afford it, as well as mandatory vaccination and virus testing. Hopefully, the testing technology will improve, so biological scanning can be conducted unobtrusively and rapidly as one enters a building or a vehicle. The exercise of government authority to require vaccination and scan for viruses should be viewed as an infringement of personal freedom in the interest of collective security. It comes back to our collective responsibility, especially those of us who live in densely settled cities.

I am especially disappointed when I see police, health-care workers, and firefighters resisting vaccination since their fundamental job is to protect the public. These people put themselves in harm's way with great frequency, so why don't they see vaccination as simply another tool they possess to protect the public? Vaccine resistance among public health and safety officials is a sad indicator of the breakdown of our sense of community. COVID-19 will not be the last global pandemic we will face; we will either combat these threats as a world community or suffer the pain and loss caused by the constant spread of disease.

COMMUNICATIONS

One of the major initiatives of the New Deal of the 1930s was rural electrification, especially the work of the Tennessee Valley Authority. Private electric utilities could not make money bringing electricity to the low-density regions of the country. But FDR and his administration decided that electricity was a necessity of modern life and government needed to provide the subsidies required to eventually attract private energy providers to rural areas. Broadband and access to the internet are the modern equivalents of rural electrification nearly a century ago. It is not always in the interest of cell-phone and internet providers to pay the cost of their communications infrastructure in areas of low population density. While progress has been made in bringing service to rural areas, over fourteen million Americans still lack access.

According to the Federal Communications Commission (FCC):

> The number of Americans living in areas without access to at least 25/3 Mbps (the Commission's current benchmark) has dropped from more than 18.1 million Americans at the end of 2018 to fewer than 14.5 million Americans at the end of 2019, a decrease of more than 20 percent. Moreover, more than three-quarters of those in newly served areas, nearly 3.7 million, are located in rural areas, bringing the number of rural Americans in areas served by at least 25/3 Mbps to nearly 83 percent. Since 2016, the number of Americans living in rural areas lacking access to 25/3 Mbps service has fallen more than 46 percent. As a result, the rural–urban divide is rapidly closing; the gap between the percentage of urban Americans and the percentage of rural Americans with access to 25/3 Mbps fixed broadband has been nearly halved, falling from 30 points at the end of 2016 to just 16 points at the end of 2019. (FCC 2021)

It's nice that the FCC focuses on progress, but in twenty-first-century America, the absence of connectivity is like lacking indoor plumbing and electricity in an earlier time. Having fourteen million disconnected people is far from trivial. In addition, we see a similar, income-based digital divide in urban areas. These are environmental sustainability issues because of the importance of the internet for modern, low-environmental-impact consumption of information, entertainment, and social engagement via electronic communication. It is also essential to be connected to the internet to be a full participant in economic life. The absence of connectivity reinforces income inequality. The digital divide in cities and rural areas is a major public policy issue. When the COVID pandemic required students to attend school online, the absence of computers and internet connection was a major impediment to education in low-income areas. This is a new form of infrastructure, but its availability is absolutely crucial.

A final communication issue that also must be addressed is the spread of disinformation for profit on digital platforms such as Facebook, Instagram, Google, TikTok, and Twitter. In a world where our very sense of reality is communicated via the internet, we must develop a legal and regulatory regime that protects the truth in communication. Like many regulatory issues in our increasingly complex and technologically based world, regulating social media requires a deep, insider knowledge of how these businesses work. The possibility of regulation dominated and captured by industry is as real in this business as it is in oil drilling or jet plane construction. Nevertheless, the danger to society of disinformation is too great to ignore. Disinformation affects policy dialogue, and even the work of scientists.

According to a report published by the George Mason University Center for Climate Change Communications:

> Climate disinformation has had many negative effects. It reduces public understanding of climate change, lowers support for climate action, cancels out accurate information, polarizes the public along political lines, and reinforces climate silence–the lack of public dialogue and private conversation about climate change. Climate deniers directly impact the scientific community–and, in turn, its ability to serve the public good–by forcing climate scientists to respond to bad-faith demands and arguably causing a chilling effect pressuring scientists to underplay scientific results. (Cook et al. 2019)

Disinformation about the 2020 presidential election and former President Donald Trump's baseless claims of election fraud threaten American democracy. Disinformation about climate change and COVID-19 threaten our ability to transition to a renewable resource–based society. Sustainability requires densely settled, sustainable cities, but persistent pandemics make that impossible. If climate change is a hoax, then why bother to stop burning fossil fuels? The complexity of modern life requires more, not less scientific literacy, as well as the regulation of disinformation. The ability to create videos of elected leaders saying words that they did not and would never say is a truly terrifying threat to democracy.

HUMAN CAPITAL DEVELOPMENT FOR SUSTAINABILITY

As an educator, I need to conclude this chapter on the infrastructure of environmental sustainability by discussing the need for organizational capacity and trained sustainability professionals. We need all of those now being trained as sustainability professionals, and we will need many more in the future. We need innovation in the management of our organizations in order to create the standard practices and capacities needed for sustainability. We also need scientists and engineers to focus their attention on renewable energy, toxics, waste processing technology, energy storage, and pollution control. Young people entering the workforce are demanding that the organizations they work for pay attention to environmental, social, and governance (ESG) issues. It is not just that people expect the organizations that they work for to act with a sense of responsibility and ethics, but that sustainability management is an indicator of overall management competence.

My view is that all competent managers must be sustainability managers, and therefore, the profession of sustainability management should eventually be subsumed by the profession of management. Of course, it could happen the other way around, and the only type of management taught in business and public policy schools will be sustainability management. No manager would be considered competent if they could not read or understand a balance sheet or a financial control system. Likewise, a manager who ignores the cost of energy or waste and potential environmental liability should be considered incompetent.

The role of the sustainability professional within an organization can vary, depending on the organization and the centrality of sustainability to its organizational strategy. For some organizations, sustainability is an organizational form of greenwashing; their goal is to make the place seem more interested in environmental protection than it is. These make-believe sustainability organizations are easy to spot and important to identify. The definition of sustainability can also vary legitimately with efforts to subsume social justice, equity, or corporate social responsibility within a very general concept of "sustainability."

As I indicated earlier in this book, and repeat here for emphasis, I agree that efforts to achieve equity and organizational responsibility fit into the overall definition of organizational sustainability. My focus, and the focus of this chapter, have been on environmental sustainability, or what I have termed the physical dimensions of sustainability: energy, water, waste, material use, and environmental impacts. Issues of social justice are quite important to the world and to me personally, and are subfields of sustainability management, as is the issue of environmental sustainability. A focus on the physical world enables objective scientific indicators and analyses to play a central role in the development of sustainability metrics and in the definition of sustainability success.

Social equity, inclusion, access, and diversity, along with stakeholder-engaged governance, are part of the overall definition of sustainability management. They are important to modern organizations for several reasons. First, we are in a brain-based, service-dominated economy. Organizations must attract and nurture the best talent. If they limit their staff recruitment and promotion by race, gender, sexual orientation, or nationality, then by definition, they are limiting the talent pool from which they can recruit and benefit. Second, many bright and talented young people judge the worthiness of the organizations that are recruiting them in part on their performance on environment, diversity, and social governance issues. Therefore, the most successful and competent organizations are those that build capacity in sustainability management. Just as we need to build physical sustainability infrastructure such as mass transit and electric charging stations, we must also build organizational capacity that is capable of building and maintaining all the productive elements of our society.

If the United States is to build a sustainable, renewable resource–based economy, the politics of environmental sustainability must undergo a fundamental change. In chapter 6, I discuss how we can articulate a positive vision of an environmentally sustainable economy and promote this vision through the media, culture, and entertainment. Environmental sustainability advocates need to create and disseminate positive role models and end the discussion of enemies and evil, selfish, and greedy consumers.

WORKS CITED

Commoner, Barry. 1979. *The Politics of Energy*. New York: Knopf : distributed by Random House, 32.

Cook, John, Geoffrey Supran, Stephan Lewandowsky, Naomi Oreskes, and Ed Maibach. 2019. *America Misled: How the Fossil Fuel Industry Deliberately Misled Americans about Climate Change*. Fairfax, VA: George Mason University Center for Climate Change Communication. https://www.climatechangecommunication.org/wp-content /uploads/2019/10/America_Misled.pdf.

Fasching, Elesia, and Suparna Ray. 2022. "Solar Power Will Account for Nearly Half of New U.S. Electric Generating Capacity in 2022." U.S. Energy Information Agency (EIA). January 10, 2022. https://www.eia.gov/todayinenergy/detail.php?id=50818.

Federal Communications Commission (FCC). 2021. *Fourteenth Broadband Deployment Report*. Docket No. 20–269. Washington, DC. https://docs.fcc.gov/public/attachments /FCC-21-18A1.pdf.

Goto, Mesura. 2013. "Global Waste on Pace to Triple by 2100." World Bank, October 30, 2013. https://www.worldbank.org/en/news/feature/2013/10/30/global-waste-on-pace -to-triple.

Ramseur, Jonathan. 2018. "Wastewater Infrastructure: Overview, Funding, and Legislative Developments." Congressional Research Service. https://sgp.fas.org/crs/misc /R44963.pdf.

Root, Tik. 2021. "U.S. Is Top Contributor to Plastic Waste, Report Shows." *Washington Post*, December 1, 2021. https://www.washingtonpost.com/climate-environment/2021 /12/01/plastic-waste-ocean-us/.

Sommer, Lauren. 2021. "The Drought in the Western U.S. Is Getting Bad. Climate Change Is Making It Worse." NPR, June 9, 2021. https://www.npr.org/2021/06/09/1003424717 /the-drought-in-the-western-u-s-is-getting-bad-climate-change-is-making-it-worse.

United States Geological Survey (USGS). 2019. "What's in Your Well Water?" March 1, 2019. https://www.usgs.gov/mission-areas/water-resources/science/domestic -private-supply-wells.

6

CHANGING THE POLITICS AND COMMUNICATION OF ENVIRONMENTAL SUSTAINABILITY

BUILDING AN AMERICAN POLITICAL CONSENSUS BEHIND ENVIRONMENTAL SUSTAINABILITY

When environmental protection was a barely noticed issue on the political agenda, it was able to garner massive support from the American public. The air pollution, water pollution, and solid and toxic waste programs of the 1970s and 1980s were not partisan issues. In 1972, the Water Act was enacted over President Richard Nixon's veto, requiring over two-thirds of the House of Representatives and the Senate to override the President's decision. Public support for these laws was well over 70 percent, and the laws were crafted by a bipartisan coalition of committed legislators. What happened?

Part of what happened was the antiregulatory ideology of the Ronald Reagan era and the related rhetoric of "job-killing regulations." But even President Reagan had to walk back antiregulation moves in the Environmental Protection Agency (EPA). EPA administrator Anne Gorsuch-Buford (yes, the mother of Supreme Court Justice Neil Gorsuch) and her associate administrator for hazardous waste, Rita Lavelle, were sent packing, and the first EPA administrator, Bill Ruckelshaus, was brought back to steer the agency back to its politically moderate moorings. The United States wanted a clean environment, and Reagan was a Californian who understood the deep public support for smog control and clean water. The Reagan administration tried a radical antiregulation ideology in 1981 and 1982, but quickly learned that political support for a clean environment required a real effort at controlling pollution.

Under Administrator Ruckelshaus and his successors up until the Donald Trump administration, EPA regulation was serious and enforcement was real, but industry was given plenty of time to comply with rules, and typically only businesses that were marginal to begin with were harmed by environmental rules. But the growing role of money in politics and fierce lobbying by ideologues and industry began to paint environmental rules as antifreedom and anticapitalist. In response to the climate issue, the fossil fuel industry intensified its lobbying and propaganda onslaught with a ferocity not seen since the tobacco propaganda wars of the late twentieth century. In both cases, those industries understood the very real dangers that their products posed and that they were in an existential battle for the survival of their businesses. By the twenty-first century, environmental protection had become an ideological political issue, particularly once the climate issue began to dominate the discourse.

One additional reason for the extreme polarization over environmental regulation was the changing media landscape. In the 1970s, there was no internet, and CNN, the first cable news network, began on June 1, 1980. When network news shows began in the 1950s, they were money-losing efforts, viewed by the networks as a public service in exchange for access to public broadcast airwaves. But starting with CNN, news became financially profitable. The business model eventually led to networks like Fox News and MSNBC, which present news and opinion with an ideological slant. Political polarization turned out to be very profitable. When combined with social media and interest groups, and with the U.S. Supreme Court decision in *Citizens United*, political speech became a growing business. In this case the Supreme Court equated political contributions with political speech and under the First Amendment, regulation of such speech was highly constrained. This unleashed the floodgates and vastly increased the role of money in politics. This in turn increased polarization as cable news became dominated by opinion rather than facts delivered according to traditional journalistic standards of multiple sources and verification. Environmental issues, especially new issues like climate change and later COVID-19, became victims of polarized politics and ideological news networks and websites.

The politics business is not limited to the right wing; ideologues of all stripes have used the technology of microtargeting to identify a core audience that they can sell to advertisers. They also make direct appeals for funds. Nonprofit interest groups have used the internet to raise donations by appealing to the fear of those holding opposite but extreme views. Some interest groups have deceived less sophisticated donors into making monthly contributions to their chosen cause on their credit cards. By monetizing extreme politics, we have delegitimized compromise and consensus, which is presented by extreme groups as a violation of principles and an attack on "truth." The idea that people may disagree about some issues but still might be able to compromise is presented as evil—a product of the "deep state," the corporate establishment, or worse. By ending the search for common values and areas of agreement, we have created a toxic, polarized politics that make legislation and consensus more and more difficult. Despite these trends, the only realistic way to transition to environmental sustainability is to build a wide base of support for modernizing our economy, and within that broad framework, reduce our environmental impact. While polarization and extremism dominate politics, there are some issues that the United States can still unite on. The nation's response to the Russian invasion of Ukraine in 2022 provides an example of our ability to act in a unified manner. When natural disasters take place, there is little question of the need for government to help people survive and recover.

In the early days of climate politics, the issue had little political salience, in part because it was very different from traditional environmental politics. The climate issue was largely ignored. Despite the machinations of politicos in Washington, broad, grassroots support for a clean environment persisted. This was because air and water pollution could be seen and smelled. The causes and effects of pollution were local and impossible to ignore. In addition, rural people who hunted and fished understood that the natural resources that they loved and depended on were in danger. In contrast, in the early days of climate politics, we saw no local climate impacts. Scientists told us that climate change was created everywhere, and its impact was in the future. We had to place our trust in, of all things, academic climate modelers and governmental earth system scientists.

But while climate policy proved problematic, other trends reinforced the importance of environmental policy. People began to focus on wellness, diet, exercise, and overall health, particularly when it came to their children. "Parenting" became an active verb (as opposed to the status of being a parent). If you didn't want your child to eat food processed with sketchy chemicals, you didn't want them breathing polluted air or drinking water with lead in it either. The "not in my backyard" (NIMBY) syndrome developed, in part, as a way of trying to prevent further real estate development and maintain local environmental quality. And then, over the past decade, extreme weather events began to accelerate and intensify, and the early climate models proved to be all too true. All the impacts that early climate models predicted began to appear on the warming planet. In recent years, young conservatives have begun to accept the science of climate change, while still rejecting the solutions proposed by progressive climate activists.

Public opinion on climate policy has been evolving, as indicated by a study by the Pew Research Center:

> While there is strong consensus among Democrats (90 percent, including independents who lean to the Democratic Party) on the need for more government efforts to reduce the effects of climate change, Republican views are divided along ideological, generational and gender lines. A majority of moderate or liberal Republicans (65 percent, including GOP-leaning independents) say the federal government is doing too little to reduce the effects of climate change. In contrast, only about one-quarter of conservative Republicans (24 percent) say the same, while about half (48 percent) think the government is doing about the right amount and another 26 percent say it is doing too much. There are also divides by age. Among younger Republicans—adults in the Millennial generation and Generation Z, ages 18 to 38 in 2019—52 percent think the government is doing too little on climate. By comparison, 41 percent among Generation X and 31 percent of Baby Boomer and older Americans say this. Republican women (46 percent) also are more inclined than GOP men (34 percent) to think the government's efforts on climate are insufficient. (Funk and Hefferon 2019)

The age measures indicate that over time, support for climate policy may grow and become less partisan. The presence of greater local impacts from extreme weather is also starting to reinforce the reality of the climate crisis.

We live on a planet that is far more crowded today than the one that some of us experienced when the EPA was created in 1970. Back then, the global population was about four billion; today, it is over eight billion. Currently, the political pressure to maintain wealth in the developed world and to build wealth in the developing world is fierce. The best way to ensure that that is done is to modernize our economies in the developed world and move toward a circular, renewable resource–based economy. To do that, we need to develop and implement new sources of renewable energy and make our electric grid capable of sending and receiving energy and operating at higher levels of efficiency. We also need to develop systems to automatically separate garbage and mine it for resources that can be reprocessed. Sewage treatment must also advance such that sewage sludge can be recycled. These high-tech solutions require additional research and development, as well as massive investment in public infrastructure.

But they hold the promise of a more productive and lower-cost economy. Energy is a growing household expense that can be reduced by lower-priced and more efficient solar cells and batteries. Electric vehicles are already demonstrating their high-tech appeal. Cities like New York are spending billions of dollars to remove garbage and send it away. In New York's case, the garbage is often sent to waste sites in Virginia and farther south. What if garbage could actually generate revenue by providing raw materials for repurposing? What if those resources were lower priced than raw materials mined from the planet? We are already seeing this in one industry. J. B. Straubel, a cofounder of Tesla, recently started Redwood, a company that makes electric car batteries in part from recycled materials. According to one article, while Straubel's company will need to mine raw materials to meet his production targets,

[t]he company's target of 100 GWh in 2025 means it can no longer rely on recycled materials alone. Unlike some consumer electronics, there's

a long lag between when electric cars are made and when their batteries are ready to be recycled. The reuse of packs in secondary applications can delay that further. Today, electric cars account for less than 10 percent of Redwood's recycling stock. "We're going to push the recycled percent as high as possible, but that is really going to be dependent on the availability of recycled materials," Straubel said. "If we end up consuming 50 percent or more of virgin raw materials, that's fine." In the decades to come, Straubel is confident that recycled materials will be used for "close to 100 percent" of the world's battery production. Recycling is already profitable, he said, and eventually companies that don't integrate recycling with refining and production won't be able to compete on cost. (Randall 2021)

In other words, some raw materials are so valuable that recycling makes economic sense. As noted earlier, we have seen this in the aluminum business, with recycled aluminum using less energy and cheaper raw materials than that made from mined resources. These new technologies for waste and energy should enable us to transition to environmental sustainability while continuing to build a vibrant, growing economy.

To build a broad-based consensus behind environmental sustainability politics, we need the basic idea pioneered by Mike Bloomberg when, as New York City's mayor, he led the development of the city's first sustainability plan: PlaNYC 2030. That plan tied environmental sustainability to economic development. In some measure, we are seeing the same impulse in the environmental elements of Joe Biden's infrastructure law and his original "Build Back Better" proposal. It's an effort to modernize the economy. A focus on building the economy, increasing employment, and developing cleaner, less expensive energy has broad, nonideological appeal. The popularity of some elements of Biden's plan stands in contrast to the bitter partisanship in Washington, which is now reflected in many communities as a place where all politics is a zero-sum game. Political opponents are now seen as bad and evil. If Biden gets something approved, even if it's something that everyone favors, his opponents see it as a political loss.

Extreme, right-wing, even Trumpian political warfare delegitimizes the political center and any form of political consensus. Any congressional Republicans negotiating compromises risk being primaried by Trumpian ideologues (and they frequently lose—and some don't even try to run again). On the left, we see environmentalists branding industry as evil and arguing that the only solution to climate change is to tax carbon and live without some forms of consumption that the public values. Politics seems to be moving toward increased polarization.

The political scientist Norman Ornstein observed:

> In the 1970s, voters tended to split tickets between, say, congressional and presidential races; in that decade the correlation of the two votes was 0.54. In the 1980s it rose to 0.65. Now it is 0.97! Just as stunning, another researcher, the political scientist Corwin Smidt, found that today's self-proclaimed independents "vote more predictably for one party over another than yesteryear's partisans." The key here, however, is what Abramowitz and Webster call "negative partisanship"—that people are now more motivated by their antipathy for the other party than by affinity for their own. The willingness of ardent Trump supporters to stick with him through scandal, outrage and actions that may damage their own economic standing becomes more understandable—criticizing him or seeing him suffer a defeat means that the evil enemy has gained a victory. (Ornstein 2020)

I recognize that building consensus in our poisonous, dysfunctional political environment may not be seen as realistic. But the only approach that seems feasible to me is to remove environmental sustainability from the ideological wars and seek the elements of change that can attract the broadest constituency. Politics seems unreal, but reality is still reality. The forest fires in the West, droughts, tornadoes, and floods in the Midwest, and extreme weather everywhere remind us that the issues of environmental sustainability are real. We all breathe the same air. We drink the same water. The food that we eat comes from the same system of industrial agriculture. The facts of our environmental condition are not based on beliefs or values, but rather objective conditions that we all experience. We are also in a global economy, competing with organizations from many nations.

The argument that we need to ensure that our energy and transportation systems are up to date is strong when it is based on the need to remain competitive. Russia's horrific invasion of Ukraine in 2022 reminded many of the nonclimate-related dangers posed by our dependence on fossil fuels. Therefore, the seeds of consensus can be found in our objective environmental, security, and economic conditions. We don't need foreign raw materials if we can mine them from garbage. Renewable energy can prevent climate change, but it also can be delivered cheaper than fossil fuel energy. Electric cars are fast becoming fashionable. Economic modernization centered in the private sector but subsidized by government-funded infrastructure and scientific research has become as American as apple pie. Economic modernization is how we can (and hopefully will) build an American political consensus behind environmental sustainability.

ARTICULATING A POSITIVE VISION
OF AN ENVIRONMENTALLY SUSTAINABLE LIFESTYLE

When people despair about our climate future and the environmental sustainability of the modern economy, I like to point out that we are in the early stages of a major transformation in how our economy operates. Evidence of change is everywhere. Hertz ordered 100,000 Teslas in 2021. More and more companies are setting pollution reduction targets. New York City's government is working to reduce the carbon footprint of its 4,000 buildings. Even more important is when these decisions slowly become reflected in facility and machinery design parameters. Engineers are asked to reduce the use of energy and toxic substances. Farmers are trying to reduce pollution runoff from their fields. Not everywhere, not all at once, but a paradigm shift has begun. A new understanding of how the world works focuses attention on reducing environmental impacts.

As economic life is changing, so are cities. In the United States, as noted earlier, 80 percent of the gross domestic product (GDP) is now in the service sector. Automation is likely to ensure that that trend continues. Machines will make things, and people will design and sell them. A century ago, most of our time was spent pursuing food, clothing, water, and shelter. Today, most of our time is spent producing and consuming ideas and information and sharing what we learn on social media. We are paying more attention to the quality and health impacts of what we eat, breathe, drink, and do. Environmental quality is now an assumed output of a well-governed modern city. As our cities become more crowded, we seek more exciting public spaces, better restaurants, better mass transportation, and better places to walk, and with assurances that the air is relatively clean.

BUILDING CONSENSUS AND CULTIVATING FRIENDS AND ALLIES

One of the victims of political polarization has been the tendency of environmental activists to shame people whose values and lifestyles they don't share. From people who eat (and even hunt for) meat to billionaires traveling to the Conference of the Parties environmental meeting in private planes—all are subject to social media shaming and criticism. Young people complain that their elders have ruined the planet. Activists argue that capitalism is greedy and a force of destruction. In my view, none of this is helpful.

As noted previously, polarization fits the business model of narrowcasting. We saw this with Fox News on the right and then MSNBC on the left. But these cable stations reach far broader audiences than many social media sites, and some organizational social media are specifically designed to reach a small substrata of the population. As noted earlier, once you identify and reach your target audience, then you can monetize them. Reaching a large, general audience with an appeal to commonly held values is far more difficult to achieve than a narrow, targeted approach. As the media echo chamber persists, people forget what they have in common and focus only on how they differ. And yet a close look at American values, and in fact global human values, shows many areas of commonality. President John F. Kennedy said about half a century ago in a commencement address at American University that some have called his "peace speech":

> So, let us not be blind to our differences—but let us also direct attention to our common interests and to the means by which those differences can be resolved. And if we cannot end now our differences, at least we can help make the world safe for diversity. For, in the final analysis, our

most basic common link is that we all inhabit this small planet. We all breathe the same air. We all cherish our children's future. And we are all mortal. (Kennedy 1963)

We all depend on this small, fragile planet and cherish our family and loved ones. This is as close as we might come to a value shared by all. And when people are in pain and in trouble, other people typically help first and conduct background checks later. In the summer of 2022, I saw a news report of a New York City taxicab that went out of control and landed on top of two women. Over a dozen bystanders ran to the cab and lifted it off the women, saving their lives. Humans are capable of collective, positive actions and achievements. We need a politics that reflects those common values.

Shaming people who do not live as you do or who view consumption differently is a good way to alienate a potential ally. A parent with several children may believe that the family needs a sport utility vehicle (SUV) so their children can be transported to activities and to see Grandma. A fifteen-year-old climate zealot may not understand the cross-pressures that that parent is experiencing. A billionaire flying a private jet to a climate meeting didn't become that rich from altruism, but they could be a powerful ally to sustainability efforts. Attacking the behavior of rich people may make advocates feel superior, but what is gained by that feeling?

Saving the planet will require a broad political consensus. The global economic system and the system of international relations are complex and at certain points fragile. The transition to environmental sustainability will not be instantaneous, and the key to its success is convincing everyone that it is necessary. It is not simply climate change, but the management of crucial resources like water and land, so their quality is not degraded. We are experiencing plastic contamination of the oceans, deforestation, destruction of ecosystems and biodiversity, and unchecked spread of invasive species and viruses. We need to build awareness and understanding of the problems, along with proposals for economically viable solutions. We need to replace fear and despair with confidence and hope. We have solved many problems in the past, and there is no reason to believe that we can't address these as well.

While wars and conflicts such as the brutal war in Syria and the equally brutal invasion of Ukraine in 2022 may cause us to doubt our ability to survive, periods of peace are far longer than periods of war in human history. Despite many wars fought since the invention of the atomic bomb, after it was used twice against Japan in World War II, it has never been used again. Despite extremism everywhere, there is a global consensus that nuclear warfare would be a lose-lose, and it is certainly true that after a nuclear war, the living would envy the dead. This is not to say that a nonstate terrorist or a crazy national leader like Vladimir Putin won't use such a weapon someday—just that there is a wide consensus against its use.

The goal should be to identify the values that we share, not those that we do not share. Hunters experience and value nature. That is a value to connect with. Athletes and parents share a concern for physical fitness and wellness. That is another value to connect with. During his 2020 campaign for president, Joe Biden often said that whenever he heard the word "climate," he thought of the jobs that combating climate change would create. He was working to connect labor to environmental sustainability.

FOCUS ON THE POSITIVE ADVANTAGES OF A SUSTAINABLE LIFESTYLE

The austere negative view of living sustainably needs to be replaced by a positive view. Walking, jogging, gardening, volunteering in parks and soup kitchens. Social engagement. Sidewalk cafés in springtime. Hiking in parks. Camping in a national park. Fishing, hunting, surfing, and enjoying the outdoors. Playing ball in a local park. Bringing a child to a playground. Visiting museums, galleries, attending concerts, going to plays and movies. Eating in a local restaurant. Inviting guests into your home for a meal and conversation. Attending a religious service or festival. Hanging out with friends watching a ball game in a bar. Gathering with family and friends for a cookout in a yard or a park. Watching a movie via streaming video in your home. Visiting a library, reading, playing a video game, listening to music on your ear buds. Attending a public meeting on a critical local issue. Calling Grandma. All these activities and many more are fully compatible with environmental sustainability. They should be celebrated and portrayed as elements of a sustainable lifestyle. These are the normal activities of American life. They don't stop because we want to reduce their impact on the planet. Our goal is to preserve this way of life by reducing, but not eliminating, its impact on the environment.

I see the impact of an exciting urban lifestyle in my neighborhood every day when I pass Tom's Restaurant on Broadway, which *Seinfeld* immortalized for all time through the iconic exterior shot of its sign. The show ran from 1989 to 1998 and has been in reruns ever since, and to this day, tourists take pictures of their families in front of the place. The collaboration of the National Aeronautics and Space Administration (NASA) with Columbia University that designed an important climate model is in the building above Tom's, and the restaurant is also featured in Suzanne Vega and DNA's 1990 song "Tom's Diner." Young people all over the world are attracted to the image of an exciting, urban lifestyle represented by television shows, popular music, and movies that take place in New York City. The place where I live and many other urban places are projected in mass and social media and have worldwide appeal.

We need artists and entertainers to create positive images of people living lifestyles that are built on concepts of environmental sustainability. Movie and television characters who visit urban parks, bike to work, explore the countryside, focus on fitness and wellness, and learn about the science of climate change and biodiversity do all of that while having a good time and going through all the trials and tribulations of normal life.

We need role models who don't spend their time complaining about evil polluters, but rather start community gardens and educate kids about the natural world. We need people who install solar panels, sell electric vehicles, or work for a business that designs and sells sustainable fashion. These role models must be seen by people around the world and create images that compete with the lifestyles of the rich and famous and others promoting greed and conspicuous consumption.

IMAGES FOR CONSERVATIVE SUSTAINABILITY SUPPORTERS

In addition to images that those of us raised in Brooklyn can relate to, we need images that some of my college classmates in Franklin, Indiana, might connect with. People enjoy outdoor life that includes hunting, fishing, hiking, and sports; whose lifestyle is centered around family and religion. People living outside major cities have seen their communities ravaged by floods, tornadoes, and other extreme weather events. They are seeing the impact of environmental degradation firsthand. They may well prefer market solutions to government intervention, but environmental sustainability requires both.

A study published in 2021 found positive results by exposing conservative respondents to a different kind of climate rhetoric:

> For example, one video features a retired Air Force general who explains that climate change poses a national security threat and creates challenges for the U.S. military. In another video, Dr. Katharine Hayhoe, a climate scientist and evangelical Christian, speaks about the consistency between her faith and caring about climate change. In another, former Rep. Bob Inglis, R-S.C., describes how his conservative values motivate his drive for political action on climate change. . . . After the campaign, the researchers compared 1,600 surveys administered before and after the campaign, which revealed that the videos increased understanding among Republicans in the two districts on two topics: that global warming is happening and that it's being "caused mostly by human activities." The understanding increased by several percentage points, according to the study. (Jacobo 2021)

The largest American environmental group has long been the National Wildlife Federation (NWF), which was started in 1936 by hunters and anglers seeking to preserve the environment and wildlife to enable them to continue to pursue the traditions they value. The environmental movement has long been a coalition between urban environmentalists and rural conservationists, and building on that tradition is one way to remove the ideological polarization that we see in today's politics around climate change.

A key element of this is a reverence for nature itself. The concluding section of this chapter focuses on building environmental values and ethics, but as the NWF demonstrates, there are deep traditions connected to environmental conservation. You can see this in the television show *Last Man Standing*, about a conservative manager of a Western sporting goods chain and his mostly liberal family. In this series, exposure to nature and a deep concern for its preservation are hardwired into the culture. Farmers, hunters, and anglers are all people who live close to nature, and in some cases they make their living from it. These are natural allies on a band of issues that urban advocates of environmental sustainability would be wise to connect with. There will be many areas of disagreement, but these two branches of the environmental movement could learn from each other.

BUILDING ENVIRONMENTAL VALUES AND ETHICS

As bad as things are in this era of COVID-19, climate, Russian war crimes, and partisan political dysfunction, I confess that I remain an optimist and believe that we will figure out the crises we now confront, and we will make the world better than it is today. COVID-19, climate, equity, racism, terrorism, war, and poverty are real and daunting public policy problems. There are people in the world that want to kidnap governors, kill a man with a relentless knee to his neck, invade neighboring nations, party during a pandemic without masks, and don't think that COVID-19 and climate change are real. But most people see the world as it is, and I find people generous and typically willing to help those in need. A growing number of people worry about our planet becoming contaminated and want to ensure that their behavior doesn't make things worse.

The importance of protecting our air, land, and water is a shared value. Polluters develop elaborate excuses and rationalizations to defend their pollution because they know that harming the environment is a bad thing, and most people see the world that way. The polluters themselves see the world that way—they can't help sharing those values. There is this nagging feeling in the back of our minds that the world is getting more crowded and the resources we once relied on are not always available anymore. The well that we dug when we first moved into our home is now contaminated, and we need to pay to pipe in filtered city water. The quiet country road that we drove on when we were kids is now a highway. The woods where we used to camp were ripped out to build a strip mall. The NIMBY syndrome comes from a desire to preserve current land uses and prevent new ones that might change a status quo that we are often eager to maintain. We are told that there is a trade-off between economic wealth and environmental protection, but if there is such a trade-off, we don't feel good about it. I should note that I consider this a false choice, and economic development that damages the environment brings short-term benefits, at the expense of much greater long-term costs.

According to a recent study published in *Nature Communications*, the costs of ignoring emerging climate crises far exceed the costs of resolving them:

> Results show that following the current emissions reduction efforts, the whole world would experience a washout of benefit, amounting to almost 126.68–616.12 trillion dollars until 2100 compared to 1.5°C or well below 2°C commensurate action. If countries are even unable to implement their current NDCs [Nationally Determined Contributions to carbon reduction], the whole world would lose more benefit, almost 149.78–791.98 trillion dollars until 2100. On the contrary, all countries will be able to have a significant positive cumulative net income before 2100 if they follow the self-preservation strategy. (Wei et al. 2020)

And as I constantly mention in this chapter, the costs can be avoided with ingenuity, scientific analysis, and carefully considered actions.

We don't want to be regulated and told what to do, but we also don't want to contaminate the air, land, and water that we all rely on. We don't want to do without modern conveniences like cars, jets, and air conditioners, but we sure wish that we could have that stuff without damaging our planet. The ethical value of environmental protection is one that people widely share.

At the start of the environmental movement in the 1960s, the issues were easier to understand. Residents of Los Angeles could see and smell the smog. Orange rivers that caught fire were obviously not clean or safe.

The toxic waste from the landfill oozing into your basement seemed like an invasion by something alien and evil.

But then the issues became more subtle and complicated. We always had forest fires, droughts, and hurricanes, but somehow climate change made them worse. Viruses were always with us, but as in the case of COVID-19, they cannot be seen or smelled. These forms of damage require the interpretation of environmental and medical experts. We have little choice but trust them to be honest and correct in their proclamations. Some people refuse to make that leap of faith, but most people know what they don't know and are willing to trust experts. We put our lives in the hands of medical doctors not because we like to do so, but because we come to realize that we have no choice. We live in a complicated, high-tech world, and we rely on experts to make it work.

We think about the issues that experts call to our attention, from COVID-19 to climate change, and it challenges us to identify behaviors that we could modify to address them. Some of these behaviors are under our control: social distancing, wearing masks, installing solar panels or light-emitting diode (LED) lights, and supporting public policies that enable us to collectively address these issues. Some are behaviors not under our control, like finding yourself standing next to a person who refuses to wear a mask and is an asymptomatic carrier of COVID-19 (unbeknownst to either you or him). People's values favor the freedom to move about freely in society and go maskless whenever they want, but their values also cause them to want to protect their loved ones from harm.

Similarly, more and more people are thinking about their carbon footprint as they turn on their air conditioners or drive their car to work. Their values could (and sometimes do) result in changed behaviors. They may look for a mass transit method of commuting, find a place to live that is closer to work, work more often from home, and purchase a more energy-efficient car and air conditioner. These values are based on a shared perception of how the world works and our current environmental conditions. It does not lead to a uniform response, but it does represent a cultural shift from the way we lived half a century ago, at the dawn of the environmental era.

Fifty years ago, no one even knew that they had a carbon footprint. Climate change and our pattern of land use development have exacerbated the impact of extreme weather on our energy system. People now routinely experience blackouts. Generator sales have increased dramatically. These objective conditions have influenced people's values, and preserving the environment continues to generate support across ideologies and political parties. However, clearly it is weaker in the Republican Party than in the Democratic Party, and that has increased the partisanship of environmental sustainability policy.

PROVIDING NATURAL EXPERIENCES FOR URBANITES

By the middle of the twenty-first century, 75 percent of the human species will live in cities. People in cities spend most of their days indoors. Virtual reality and video games now compete with sports and physical activities like hiking and skiing. A reverence for nature is an important element of the value framework for protecting the environment. At some point, we will develop the technology to cure cancer. Does that mean we should feel free to flood the planet with toxics that are carcinogenic? Humans might survive exposure, but what about other living parts of our biosphere? If virtual reality can provide us with an artificial, naturelike experience, why preserve nature? We can get our nature in a version of *Star Trek*'s holosuites, where we will be transported into a computer-generated version of it.

To preserve the planet, people will need an emotional connection to its genuine and actual features. That will require a self-conscious effort by our educational institutions and by our culture as a whole to bring city folks to the countryside and to build that connection. Urban parks also have a role to play since they can attract people outdoors. Nature experiences do not have to be constant to have a profound impact. I remember camping trips that I took fifty years ago, and even if I miss a few days of walking in the park, the experience stays in my mind and provides a sort of psychological access.

But the seductiveness of the virtual world should not be underestimated. As Joe Drape reported in the *New York Times* in December of 2021:

David, 13, and Matthew, 11, are fledgling e-sports athletes. David and Matthew are part of a surging migration among members of Generation

Z—as those born from 1997 to 2012 are often labeled—away from the basketball courts and soccer fields built for previous generations and toward the PlayStations and Xboxes of theirs. It's not a zero-sum game: Many children . . . enjoy sports both virtual and physical. But it's clear that the rise of e-sports has come at the expense of traditional youth sports, with implications for their future and for the way children grow up. (Drape 2021)

The design of a virtual experience is not bound by the limits of physical reality; it can be quite stimulating and imaginative. It can also be designed to be psychologically addictive. These virtual experiences compete with the real world, and unless young people are taught to value nature, virtuality will often win over reality. The battle is over the use of time. Quiet contemplation of a beautiful vista competes for views with a stimulating, rapidly changing set of virtual images and web-based information. To some of us, this seems like a science fiction scenario, but it is quite real. If reality can be manipulated, it can be replaced unless we infuse the value of genuine and authentic experiences over those that can be created.

It will be up to families, along with educational and religious institutions, to teach an appreciation and even reverence for nature, starting with prekindergarten and extending through higher education. In the United States, this will be particularly important since we are already developing a highly polarized society, divided culturally and politically between large coastal cities and smaller towns and rural areas away from the coasts. An irony of communications technology and virtual reality is that it has the ability to transmit different notions of reality to different subgroups. Real-world exposure to nature can counter those images and expose people to each other.

In addition to bringing city folks to the country, it will be important to provide outdoor experiences closer to home as part of daily life. In PlaNYC 2030, one goal was to ensure that every New Yorker was within a ten-minute walk from a park. Nature, recreation, noncommercial human engagement, and communal and democratizing spaces are important to creating a shared experience that can take people away from their subgroup and into a collective resource. I am fortunate to live across the street from Morningside Park in New York City, three blocks from Riverside

Park and about fifteen blocks from Central Park. I am able to experience changing seasons; see every breed of dog that exists; watch children screaming with glee in playgrounds; observe personal trainers at work; cheer Little League baseball, soccer, and world-class skateboarding; and listen to live music in the open air. We need to ensure that cities provide alternatives to screen time and places where people can get away from the social media echo chamber.

CONCLUSION: CHANGING THE POLITICS AND COMMUNICATION OF ENVIRONMENTAL SUSTAINABILITY

There are a large number of factors contributing to political polarization, and others that challenge the values and ethics of environmental protection. Currently, we do not have the technology to supplant nature. We can't geoengineer a solution to global warming. We can't rapidly and efficiently overcome a virus like COVID-19. Our understanding of our planet and our species is growing, but we still have a lot to learn. But someday we will learn how to do without nature. Is that knowledge that we want to use? Is that a technology that we want to deploy?

We invented the atomic bomb, used it twice, and decided to set it aside. As close as we came to a nuclear exchange during the Cold War, we somehow managed to avoid it. While many feared that Russia's war with Ukraine might turn nuclear, even that conflict has avoided the use of nuclear weapons thus far. Similarly, we might create a taboo around our technological ability at some point in the future to supplant nature.

In my view, the politics of environmental sustainability must move away from polarization and toward consensus. There is a firm basis for consensus in the objective degradation of our environment and our need to reverse that degradation. The alternative is a world that we would not recognize where technology has replaced nature. While many Americans live in enclaves that appear unaffected by environmental degradation, we are all one storm away from experiencing firsthand the impact of global warming. Many of us have been touched by COVID-19, and others will confront similar threats to our well-being. These objective conditions are widespread and unpredictable. Of course, the same can be said about gun violence and the ideology of Second Amendment advocates that enables people to avoid connecting mass and individual shootings to the gun lobby. And therefore, while there is a firm basis for consensus, there are certainly economic and ideological interests available to push polarization.

The case that I am building here is designed to influence climate activists and advocates of environmental sustainability. Polarization politics is a trap that we have no time for. Symbolic gestures have the nutritional value of cotton candy at a time when we need substance and protein. We are in a race against time, and as I discuss in the conclusion of this book, the transition to environmental sustainability is a generations-long struggle. We need to take a realistic look at past transitions to see what made them work. When we discuss the Green New Deal, we need to look at FDR's New Deal and all of its politically induced compromises and shortcomings. Polarization politics turns perfection into the enemy of the good. The search for common ground is the key to building a consensus behind environmental sustainability. A stronger, more resilient economy with lower-priced energy and cleaner air and water is difficult to argue against. The last time I looked, everyone likes to breathe and no one wants to take poison. Let's build a consensus around the still-deep support for a clean environment and articulate a positive vision of environmental sustainability.

WORKS CITED

Drape, Joe. 2021. "Step Aside, LeBron and Dak, and Make Room for Banjo and Kazooie." *New York Times*, December 19, 2021. https://www.nytimes.com/2021/12/19/sports /esports-fans-leagues-games.html?searchResultPosition=8.

Funk, Carly, and Meg Hefferon. 2019. "U.S. Public Views on Climate and Energy." Pew Research Center, November 25, 2019. https://www.pewresearch.org/science/2019/11 /25/u-s-public-views-on-climate-and-energy/?utm_source=adaptivemailer&utm _medium=email&utm_campaign=19-11-25%20climate&org=982&lvl=100&ite =5010&lea=1139465&ctr=0&par=1&trk=.

Jacobo, Julia. 2021. "Tailoring Climate Change Messaging for Conservatives Could Shift Understanding of Crisis: Study." ABC News, June 14, 2021. https://abcnews.go.com /US/tailoring-climate-change-messaging-conservatives-shift-understanding-crisis /story?id=78257394.

Kennedy, John. 1963. "Commencement Address at American University." https://www
.jfklibrary.org/archives/other-resources/john-f-kennedy-speeches/american
-university-19630610.

Ornstein, Norman. 2020. "Why America's Political Divisions Will Only Get Worse."
New York Times, January 28, 2020. https://www.nytimes.com/2020/01/28/books
/review/why-were-polarized-ezra-klein.html.

Randall, Tom. 2021. "Tesla Co-founder Has a Plan to Become King of EV Battery
Materials—in the U.S." *Fortune*, September 14, 2021. https://fortune.com/2021/09/14
/tesla-cofounder-jb-straubel-redwood-materials-battery-materials/.

Wei, Yi-Ming, Rong Han, Ce Wang, Biying Yu, Qiao-Mei Liang, Xiao-Chen Yuan,
et al. 2020. "Self-Preservation Strategy for Approaching Global Warming Targets
in the Post-Paris Agreement Era." *Nature Communications* 11 (1): 1624. https://
doi.org/10.1038/s41467-020-15453-z.

CONCLUSION

The Long Transition to Environmental Sustainability
Is Already Underway

THE EVOLUTION FROM A MANUFACTURING
TO A SERVICE ECONOMY

While the transition from today's service and brain-based economy to an
environmentally sustainable economy will take decades to complete, the
change has already begun. Earlier in this book, I outlined the phases of
the transition. The time scale of the change is important to understand.
I base my sense of the time scale on a change process that I, along with
many other residents of New York City, unknowingly lived through. It was
New York's painful evolution from a manufacturing city to a city with a
radically different economic base. In the years between the end of World
War II and 1980, New York lost over one million manufacturing jobs. In an
analysis of the city's "apparel cluster" by Harvard graduate students Cassie
Collier, Helena Fruscio, Helen Lee, and Janet Tan, they wrote:

> Once a city producing 95 percent of the country's apparel, NYC—due
> to the high cost of doing business, space constraints, and availability of
> cheaper options overseas—now produces just 3 percent of the total cloth-
> ing in the US. (Collier et al. 2015)

In the late 1970s, these job losses started to be countered by job gains in
the service sector. According to a *New York Times* piece published in 1981:

> In a survey of factory employment, the [U.S. Regional] commissioner,
> [of Labor Statistics] Samuel M. Ehrenhalt, noted that New York City had

lost more than half of the 1,073,000 manufacturing jobs of 1947. . . . In 1980, manufacturing employment in New York City decreased by 20,000, to 499,000, bringing the total decline for the last three years to 40,000. While these losses were taking place in manufacturing, Mr. Ehrenhalt said, there was an increase of more than 100,000 in service jobs and also advances in finance, insurance and real estate, and in contract construction that as a group, provided the basis for the city's 1977–80 job recovery. (*New York Times* 1981)

The economic transformation during the 1960s and early 1970s was characterized by job losses and the near-bankruptcy of New York City in 1974. Through the 1980s and 1990s, we began to see the replacement of manufacturing by service businesses such as finance, media, information, health care, education, consulting, and insurance. In the twenty-first century, social media and e-commerce companies arrived, along with a growing number of technology start-ups.

The forces behind the change included lower-priced labor elsewhere, but also technological change. The city's West Side docks were too small for containerized shipping, and factories became more automated and required large one-story spaces for assembly lines. Until the 1960s, New York manufactured clothing, bicycles, cars, and appliances. Today, manufacturing is a small part of New York's economy. The change in the city's economy was unplanned and coincided with civil unrest, increased crime, and outmigration to the local suburbs and other states. The decline was seen at the time as a trend that would not be reversed. But abandoned loft factories in SoHo became artist studios, and then luxury homes and spaces that once housed factories found other uses.

By the first decade of the twenty-first century, New York City had been revived as the most international city in the nation, as well as a gateway to the global economy. The city's population began to grow. In 1950, New York City had about 7.9 million people. By 1980, the population had declined to a little more than 7 million, and in 2000, it reached 8 million. In 2016, it peaked at 8.5 million, dropping to about 8.3 million in 2021. Post COVID, New York has resumed growing in population with current estimates ranging from 8.5 million to 8.9 million. As the workforce becomes more hybrid, mixing office work with work at home, New York's

Central Business district will become more residential and its residential neighborhoods will become more commercial. This can happen because the city's economic function has changed dramatically over the past seventy years.

In fact, New York's role has changed from apparel manufacturer to a key center of the global fashion industry. This was noted by those Harvard students I cited earlier: Collier, Fruscio, Lee, and Tan. According to their analysis:

> Though NYC no longer competes in large-lot apparel manufacturing, the city has a strong competitive advantage in sample-making and small-lot production, defined as production of less than 50 pieces. . . . Because the NYC apparel cluster is close to fashion consumers, industry players can test out new styles, get fast feedback, and quickly iterate on the design. This quick turnaround is possible thanks to NYC's highly-concentrated network of apparel experts that allows a designer, for example, to draw a sketch of a dress and walk it to the office across the street to a skilled sample-maker who can bring the design to life in a matter of days. The speed and flexibility that NYC offers in serving smaller markets of highly-differentiated and higher-priced fashion product is a key strength of the cluster. . . . NYC stands out globally for its concentration of wholesale and showroom establishments. Each year, about 578,000 individual wholesale buyers and fashion event attendees visit NYC, with the wholesale market alone contributing $16.2 billion in direct spending annually. The Javits Center is a particularly important place, as it is the hub for a series of trade shows and holds approximately 5,000 showrooms. The effectiveness of the region as a place for convening is evident in NYC's re-visit rate. The . . . average buyer visits NYC 4.2 times a year to attend fashion trade shows. (Collier et al. 2015)

These changes in the city's economy did not go unnoticed by people who were winners and losers during the transition, but the degree of transformation was not well understood by the public or the city's elected leaders. In the post-COVID environment, most of New York City's face-to-face advantages began to reassert themselves, even in a work environment that has more online or hybrid elements.

PROGRESS TOWARD SUSTAINABILITY BY STATE AND LOCAL GOVERNMENTS, CORPORATIONS, AND IN FINANCE

Government

This transition in fashion and in the overall economy of New York City went by relatively unnoticed until it was nearly completed. New York City's sustainability initiatives are also underway, but they also have flown under the radar. The growth of bike sharing and bike lanes is one initiative. Grading of large buildings by the New York City government for energy efficiency is another. Other environmental sustainability initiatives include the following:

- A ban on natural gas hookups in new buildings
- An electric heat pilot project in two New York City Housing Authority (NYCHA) developments
- Congestion pricing to begin by 2024 in Manhattan's Central Business District
- An effort to decarbonize the city's own buildings and vehicles
- Movement of utility rooms in many buildings from basements to higher floors, in response to sea-level rise
- Major initiatives by electric utilities with small businesses and home-owners to save electricity, funded by a state tax on electricity
- A wind farm being constructed off the south shore of Long Island

There are scores of other initiatives in New York and thousands all over the country. We see windmill construction in Texas and solar farm construction in California, efforts to recycle Christmas trees and compost food waste in many cities, and construction of urban parks on old industrial sites from Portland, Oregon to Washington, D.C.

Corporations

Many private companies have started the process of examining their production process supply chains for waste, toxics, energy use, water use, and greenhouse gas emissions. According to the Governance and

Accountability Institute, corporate sustainability reporting is increasing each year. In its 2020 sustainability reporting analysis, it observed:

> G&A's annual research series began nine years ago with the analysis of sustainability reporting activities for publication year 2011, when we found just about 20 percent of the S&P 500 companies were publishing a *sustainability* report. G&A has found the volume of reporting has steadily increased each year since 2011 and the contents of the reports dramatically expanded over time. By 2012, more than half (53 percent) of the companies were publishing reports. That percentage grew to 75 percent by 2014 and to 86 percent by 2018. In the just-completed 2020 research, examining 2019 reporting activities, G&A determined that 90 percent— 9-of-10 companies in the S&P 500 were publishing a sustainability report. (Peterson 2020, emphasis in original)

The increase in companies reporting over the past decade or so is remarkable and a strong indicator of the growing awareness of the importance of sustainability management. It is likely that some of the change that we are seeing is superficial, and it is obvious that a learning process is underway. Reporting is in response to investor demand and may also reflect internal organizational pressure brought by younger workers. However, the 2022 proposal by the Securities and Exchange Commission (SEC), referred to earlier mandating and defining climate reporting, signals a new and more rigorous approach to corporate sustainability reporting. Climate disclosure will not be voluntary and run by NGOs but mandatory and, like financial accounting, regulated by the U.S. federal government.

The movement of the American economy to the service sector and increased automation in manufacturing means that less and less employment and wealth are tied to making physical objects and more are about design, communication, information, and services such as health care, education, hospitality, entertainment, and the arts. These changes provide opportunities for reengineering both manufacturing and service delivery. Technologies like electric vehicles and renewable energy provide opportunities to decarbonize elements of e-commerce. Automation can help stimulate local manufacturing in new facilities that can design production processes according to the principles of industrial ecology.

New services such as physical training and other wellness services such as yoga and group exercise classes are common, along with service-oriented professions including event planning, home organizing, furniture assembly, "junk" removal, home improvement, elder and child care, home communications, and computing and entertainment installation and upkeep. These growing areas provide opportunities to redesign service delivery to reflect environmental sustainability priorities. It is clear that environmental issues are taken seriously by many service providers as they design their production processes. They are in the culture, and consumers ask about sustainability and service providers advertise the sustainability feature of their offerings.

Issues such as the toxicity of cleaning substances, the climate footprint of services, and the disposal of electronic waste are raised by customers and require a response by service providers. This is a cultural shift that is underway and unmistakable. We don't yet know what it means and how durable these changes are. I am confident, however, that this is the start of a paradigm shift. I am equally confident that the transition to a renewable resource–based economy will take a long time.

We are all heavily invested in the technology of the linear, fossil fuel-based economy. I drive a car that has an internal combustion engine. Little of the energy in my apartment is renewable. The waste from my household is not yet mined for raw materials. Still, every day we see examples of corporations, nonprofits, and government agencies taking steps toward sustainably managing their operations. Food companies and restaurants are recycling their waste, auto companies are investing billions of dollars in electric vehicles, and real estate developers are building structures designed to use less water and energy. New businesses are starting to share products and services or recycle products. Governments are adopting green procurement principles, purchasing electric vehicles, and making their buildings energy efficient.

Many businesses are seeing growth opportunities in the green economy. New technologies, new services, new knowledge, and new jobs are emerging. The old, clunky, pollution-belching smokestack once represented economic might in the twentieth century. The twenty-first-century version includes examples such as passive solar–designed buildings with park views, housing companies that develop smartphone applications, health care facilities, and ride-sharing services. The high value-added parts of our economy are services, ideas, and software; manufacturing is more automated and produces less wealth. Reducing the costs of energy, water, and other materials is occupying more creative energy than ever before. This drive toward a green economy is not entirely based on idealism; it has roots in a desire to be profitable while protecting the planet.

Why doesn't the reality of this movement receive much attention? In part, it is because this is good news, and people would rather watch reports of a natural disaster than of a human-made wonder. The media is a business, and its job is not to portray the world we live in, but to make money. The world that we see through the news media is not the world that we experience firsthand. The most important changes in how we live largely go unreported, while conflict, murder, and mayhem flood what we used to call the "airwaves." The day-to-day lives of families and friends generally make the news only when they involve tragedy.

The smartphone and global communications are probably the technologies that have had the most impact on human behavior of any invention in the last fifty years. People spend more time than ever communicating with each other and sharing facts, events, photos, and perceptions. This is not newsworthy, but like the move toward renewable energy, it is a central part of our contemporary reality. Unfortunately, the media is an unreliable source for understanding the world that we live in. News reports must be analyzed for their bias, especially why a particular story manages to make its way onto a crowded media menu. The progress that we are making toward sustainability is not always easy to see and is often contradicted by movement away from sustainability. The picture is complicated and sometimes contradictory.

While the transition to a renewable resource–based economy is well underway, there remain plenty of unsustainable practices and businesses in the world. And the people who benefit from those practices and businesses are not shy about defending them. Moreover, the news media loves reporting on the conflict between climate advocates and fossil fuel interests. The movement away from fossil fuels is not without victims. Writing about the decline in Wyoming's coal industry and the simultaneous rise of that state's wind business, Coral Davenport of the *New York Times* observed: "The new positions and financial opportunities offered by wind and other new-energy industries are not replacing all the jobs going up in coal smoke" (Davenport 2016). Not only is the number of jobs smaller, but the skill base is different. While these transitions are inevitable in a capitalist society, government programs and policies are needed to ensure that the victims of this transition receive the help that they need to find meaningful employment.

Even among those whose jobs are not under threat, we see evidence of resistance to sustainability everywhere. For instance, opposition to complying with air pollution rules was embedded within Volkswagen's organizational culture, permitting thousands of vehicles to be sold with software designed to fool air pollution inspection tests. This resistance to good practice is far from rare. But it is rapidly becoming the exception, not the rule. Even Volkswagen found it necessary to change its leadership and focus on the development of electric vehicles. But the company's transition from polluter to electric vehicle innovator has largely gone unreported.

The transition to a renewable economy will not be instantaneous. It will be a matter of two steps forward and one step back, and for some people, it is simply coming too slowly to save the world. Unfortunately, while the current pace may not be fast enough, it will have to do. It takes people and organizations a long time to change. Humans are ingenious and creative, but we are also creatures of habit.

As new technologies are developed that are less destructive to ecosystems, their adoption may begin slowly, but it may also pick up momentum quickly once they've passed the tipping point of popular acceptance. Cell phones, global positioning systems (GPS), and Bluetooth are examples of technologies that were adopted gradually at first, but with increased speed as they became more common. We are seeing a similar phenomenon with rooftop solar cells, solar water heaters, and the electric car.

We are also seeing this in building construction with the development of more efficient, modular, factory-based construction and the net zero building. According to the advocacy group Team Zero:

> In simplest terms, a zero net energy building is one that produces as much renewable energy as it consumes each year—the "net" referring to the annual balance between energy production and energy consumption. In non-technical contexts, many practitioners are migrating to the more conversational term, "zero energy." The most common renewable energy source for a zero-energy home is a photovoltaic (PV) array, typically roof-mounted but occasionally freestanding on the building site. However, zero net energy is achieved by working on energy reduction, as well as on energy production. Thus a high degree of energy efficiency is at the core of zero-energy projects; without that foundation, few projects would have space for a PV array large enough to meet their annual energy needs. (Edminster 2020)

A 2021 study by this group found that there were 28,000 net zero buildings in the United States and Canada. In a late 2021 *New York Times* piece on net zero buildings, Jane Margolies observed: "Demand for residences that produce as much energy as they consume is being spurred by climate concerns, consumer appetite and more affordable solar technology" (Margolies 2021).

The transition to a sustainable, renewable resource–based economy will require a more sophisticated partnership between government and the private sector than the federal government, particularly Congress, seems capable of undertaking right now. Fortunately, many city and local governments seem more adept at forging these relationships. Ideology is less important at this operational level of government, where most decision-making focuses on tangible projects and programs rather than symbolic policies and positions.

The change that we need will be given operational meaning by the organizations we work for. Just as our organizations learned to incorporate occupational safety, financial reporting, performance measurement, customer relations, employment law, social media marketing, and many other elements into their standard operating procedures, so will they need to incorporate a concern for the physical dimensions of sustainability. They will pay more attention to their use of energy, water, and other materials. They will think about recycling and reuse of finite materials, and staff will devote significant effort toward reducing the environmental impact of organizational outputs. Change will be slow and steady, and like the tortoise, may not attract much attention until it finally wins the race.

While it is not noticed much in the media, many companies are moving forward on environmental sustainability initiatives. Of the hundreds I might select, here are three that have caught my eye: Apple, Etsy, and Walmart.

APPLE

Apple's environmental strategy focuses on climate, resources, and chemistry. According to its 2021 *Environmental Progress Report*:

> Climate Change: We've set a goal to become carbon neutral across our entire footprint by 2030. We will get there by reducing our emissions by 75 percent compared to 2015, and then investing in carbon removal solutions for the remaining emissions.
>
> Resources: We aim to make products and packaging using only recycled or renewable materials. At the same time, we're committed to stewarding water resources and sending zero waste to landfill.
>
> Smarter Chemistry: Through chemistry innovation and rigorous controls, we design our products to be safe for anyone who assembles, uses, or recycles them—and to be better for the environment. (Apple 2021)

Former Environmental Protection Agency (EPA) chief Lisa Jackson heads up Apple's sustainability work, and even a casual review of Apple's website will convince you that the company is serious. Reducing the environmental impact of their products and processes is clearly imbedded into its corporate culture. Environmental impact is a design parameter when new products are built. Apple's sustainability commitment is likely a result of the relatively young age of the people who buy its products, as well as the type of employees it attracts. As in all three of the cases presented in this chapter, the absence of audited and certified sustainability metrics makes it difficult to confirm outputs and outcomes. Hopefully, the proposed SEC climate disclosure rules will finally begin the process of providing such metrics.

ETSY

Etsy is the Amazon of small, craft-oriented businesses, and it has built a brand in part out of its commitment to all the subfields of sustainability, including diversity, equity, and transparency. In the area they term "ecological impact," they have set and met ambitious goals. Chelsea Mozen recently reported on Etsy's blog that the company recognized

> [t]he urgency and importance of doing our part in trying to prevent the worst effects of climate change. That's why this year we set a long-term carbon-reduction goal that is aligned with science. Our new Net Zero by 2030 goal includes a 50 percent absolute reduction in our Scope 1 and Scope 2 greenhouse gas emissions and a 13.5 percent absolute reduction in our Scope 3 greenhouse gas emissions. And we're already meeting ambitious milestones along our carbon reduction journey. We achieved our 2020 goal of sourcing 100 percent of our electricity from renewable energy. We also reported significant energy efficiency gains in our computing with a 23 percent reduction between 2018 and 2020, despite substantial growth in our business over the same time period. (Mozen 2021)

The company also diverts 90 percent of its waste from landfills by either recycling or sending it to waste-to-energy plants, and is working to reduce the carbon footprint of its supply chain. Etsy has demonstrated leadership and commitment in other sustainability areas as well. Perhaps most notable is its work with Code Nation, a nonprofit that works to diversify the talent going into the computer industry. As Etsy's Nick Morese observed, the company demonstrated its

> commitment to fostering diversity in the tech industry by donating space in our headquarters to Code Nation, a non-profit organization that provides students from under-resourced high schools with access to essential technology, coding, and professional skills. . . . Etsy has been a proud part of Code Nation's community since 2014, hiring interns, hosting events, and empowering our engineering team to lead an on-site coding class at Etsy. We're pleased to extend that support in 2020 by donating 6,000 square feet of space in our Dumbo office for the year. (Morese 2020)

While Apple and Walmart are older organizations with origins that predate the so-called sustainability era, Etsy is a newer company, with sustainability at the center of its self-definition. It is working in every possible way to be a profitable business that acts as ethically as it can. That focus is central to both its self- and brand identities. Its commitment to sustainability attracts management, other staff, and investors who find this identity key to their own engagement with the company. I am confident that we will see many more organizations developing a similar focus in the next several years.

WALMART

Walmart, the world's largest retailer, came to environmental sustainability because it saw the potential for financial gain from renewable energy, waste reduction, and other environmental sustainability practices. Walmart has long used its vast influence in the marketplace to push its suppliers toward environmental sustainability. Its deep experience has resulted in operational involvement in the sustainability issues faced by its suppliers. In December 2021, it began a program with HSBC to provide low-cost capital to facilitate suppliers' sustainability initiatives. According to Walmart's website:

> Walmart today raised the bar on climate action by creating a supply chain finance program that not only enables greenhouse gas (GHG) emissions

reductions, but for the first time, uses science-based targets to do so in a way that aims for a 1.5-degree Celsius pathway. . . . Eligible suppliers can approach HSBC for early payment on their invoices approved by Walmart with pricing on the financing linked to the supplier's CDP scores, targets set and impact reported. Suppliers setting the highest ambition would be able to take advantage of receiving the lowest pricing. The partnership with HSBC helps Walmart address its Scope 3 emissions and supports its suppliers to reduce their Scope 1 & 2 emissions. (Walmart 2021)

This deep dive into the supply chain and the financial issues that might limit environmental sustainability is an indicator of the company's leadership in this area. Its approach to waste reduction and climate change is impressive and appears to be an integral part of its corporate culture. It is becoming increasingly common to see the roofs of Walmart's huge retail outlets covered with solar arrays. According to the company's corporate website:

According to the Solar Energy Industries Association, in 2019, Walmart added the most solar of any company in the U.S., increasing our solar use by more than 35 percent. This growth in solar was driven by several large offsite solar projects added to our long history of using solar at our facilities. And according to the EPA Green Power Partnership Top 30 Retail Ranking, Walmart was the top retailer in terms of annual green power usage in the U.S. in 2020. These recent strides have moved us closer to meeting our goals. In 2020, renewable sources supplied an estimated 36 percent of our electricity needs globally. (Vanderhelm 2021)

Cost savings due to the adoption of renewable energy is not controversial and does not require acknowledgment of the reality of climate change. Thus, Walmart could simply focus on utilizing lower-cost energy without addressing the issue. Nevertheless, the company is very direct and explicit about its understanding of climate change and its goal of reducing its carbon footprint. According to Walmart's corporate website:

As a retailer with operations in more than two dozen countries and sourcing that spans the globe, Walmart is deeply committed to addressing climate change. We're focused on strengthening business resilience, advocating for climate action and targeting zero emissions across our global operations by 2040, without relying on carbon offsets. In 2016, Walmart was also the first retailer with a science-based target designed to achieve emissions reduction in our own operations and supply chain. . . . In 2020, we raised our aspiration to reduce emissions in our operations (scopes 1 & 2) by realigning our science-based target to a 1.5-degree Celsius trajectory. . . . Because most emissions in the retail sector lie in product supply chains rather than in stores and distribution centers, we're also working with suppliers through our Project Gigaton initiative to avoid a gigaton of greenhouse gas emissions from the global value chain by 2030. . . . More than 3,100 suppliers have signed on. (Walmart 2022)

What is particularly important about Walmart's engagement with environmental sustainability is that it is not expected (or maybe even noticed) by most of its customers and staff. While being green is expected by many Etsy and Apple customers, Walmart's customer base is different. There is likely a large number of Walmart customers who are climate deniers and would say that they do not "believe in" climate change. In Walmart's case, environmental sustainability is an integral part of its business strategy. It enables the company to reduce the cost of energy and waste management and helps it make money.

A good example was an early decision of the company to sell only concentrated laundry detergent. Previously, discount manufacturers would produce large containers of detergent that looked like a bargain but were largely water. Walmart's business model is to essentially lease shelf space to suppliers, and therefore smaller, higher-priced bottles of concentrated detergent made more money per square foot of shelf space than the larger, cheaper containers filled with water. On the environment side, the plastic containers were smaller, and shipping a detergent that was mainly water required more energy and truck space to transport. Very early on, Walmart understood the efficiency savings that could be achieved with environmental sustainability.

Finance

In December 2021, Amy Myers Jaffe examined the state of sustainability finance in a terrific article entitled "This Was the Year Investors and

Businesses Put Big Bets on Climate." This was no surprise. My Columbia colleague, Professor Satyajit Bose, created and leads our certification program in sustainability finance, and he, our students, alums, and faculty engaged in this area have been working to develop and grow this field for about a decade. The piece by Jaffe presented several data points worth noting:

> Investor demand for climate-friendly stocks has surged in the past couple of years. Tesla Inc., for instance, had a market value of more than $1 trillion as of Friday, up from $300 billion in the summer of 2020. Hydrogen-fuel-cell firm Plug Power Inc. had a market capitalization of $19 billion, up from $270 million in 2018. The stock price of First Solar Inc., a U.S. integrated solar firm, was near its five-year high at $97 a share, rising steadily from barely $30 at the start of 2017. Clean-energy ETFs hit more than $25 billion in total assets in the first half of 2021, making their ownership as preponderant of a trade as technology stocks were during the dot-com craze (although I'm certainly not implying an inevitable collapse). (Jaffe 2021)

Of course, some of these increases must be viewed as part of the overall market rise in 2021. Nevertheless, not only were investors interested in these green stocks, they were also less interested in fossil fuel companies. I strongly believe that the only way for fossil fuel companies to survive is to redefine themselves as energy companies and gradually self-divest from fossil fuels. The organizational capacity of these firms should be utilized to get into the renewable energy business as quickly as possible. They might want to remember Kodak, a company that helped invent electronic photography but eventually went bankrupt because it was unable to market its inventions in photo technology and did not understand changing consumer preferences. Contrast that to AT&T, a monopoly that was broken up by the federal government but survived through multiple technological revolutions to this day because it moved from telegraphs to landline phones, and then to cell phones, defining itself as a communications company. Fossil fuel companies can lobby Congress all they want, but in the end, they will either adapt or go out of business. They don't need to believe me, but they should focus on what the financial markets are saying.

Fossil fuel companies have been losing ground to renewable energy firms in attracting new capital, as Jaffe reported:

> When it comes to investing in real, nonfinancial assets, the shift is even more stark. Private-equity firms that were previously focused on traditional oil, gas and coal assets are pivoting to green portfolios as a selling point to institutional investors, especially pension funds. The firms raised $52.2 billion for new renewable funds in 2020, up from $44.6 billion in 2019. That contrasts with conventional oil, gas and coal funds, which brought in $8.3 billion last year, down from $20.9 billion in 2019. . . . In contrast to the money being made on this climate-change mania, fossil-fuel-company stocks have fallen out of favor. Even with $70 oil, oil and gas stocks barely make up 1 percent of the S&P 500, down from 4 percent a year ago and over 11 percent a decade ago. (Jaffe 2021)

The financial movement is an indication that climate science has been accepted by smart people who have decided to bet their livelihood (or at least their clients' money) on the necessity of addressing global warming. Investors have been reflecting trends in society as a whole—they are not leading here. Pension funds and university endowments have been directed to invest in the green economy. Some are under pressure to divest from fossil fuels, but the long-term market performance of the fossil fuel companies would be driving institutional investors away anyway.

The effort to decarbonize the world economy will be a generations-long struggle, but it's very clear that 2021 represented a turning point of sorts. That was the year that we saw massive renewable energy and climate adaptation investment in President Joe Biden's federal green procurement policy and trillion-dollar bipartisan infrastructure law. Those moves marked a stark departure from the climate denialism of Donald Trump and his administration. The U.S. government has tremendous influence and can create an investment environment that reinforces market tendencies.

As a sustainability educator, I am struck by the need for professionals in this field to be educated in a number of disciplines if they are to be successful in understanding and advising investors about environmental risks and opportunities. The typical finance student in business school tends to know very little science. Occasionally, you'll see business school students with an engineering or computer science background, but they rarely know much about climate science, climate modeling, ecology, or earth science. Even people with an engineering background may not understand renewable energy and microgrids. Unlike standard investment analyses, looking at consumer demand, debt structure, revenue, earnings, return on investment, production processes, and even political risk is insufficient. Environmental risk is complex, and no single professional can understand all the necessary science to correctly advise others on investing. What we need are professionals who know what they don't know and are good at tapping into the expertise needed to evaluate environmental impacts, along with the usefulness and feasibility of emerging green technologies.

The effort to reduce greenhouse gases from energy production and consumption is only the first phase of our drive to eliminate human-induced climate change. Much more challenging work will follow. Greenhouse gases such as methane are produced in part by landfills and livestock, which need to be controlled. Carbon dioxide is produced by cement manufacturing, and other industries that emit greenhouse gases also need to be addressed. In addition, green technology is increasing demand for minerals that are being mined in destructive ways. Unlike energy generation, some of these production processes have not received a great deal of attention, but once the momentum behind green energy is established, we should expect to see increased attention paid to these critical environmental issues.

Over the next several decades, the transition to a renewable resource–based economy will require massive infusions of capital. That can be accomplished only if financial market professionals understand enough science to navigate uncertain conditions and new technologies. A clear regulatory environment is needed to assure investors that we are serious about mitigating climate change. The United States plays a critical role in leading climate policy. My hope is that even if a conservative federal government abdicates its climate leadership role once again in this country someday, the market forces launched in 2021 will be too deeply established by that time to deter the forward progress.

HOW LONG DOES TRANSITION TAKE?

I project that the transition to environmental sustainability will take place in stages. The developing world will lag the developed world, although my hope is that some developing nations will manage to avoid some of the worst environmental impacts of the development process in nations with more mature economies. In the United States, we will see the movement to environmental sustainability vary by state. A rapid transition will require public-private partnerships, which will require active state and local governments that are willing to raise and spend tax dollars, as well as energy utility commissions that are not dominated by the utilities they regulate. It will require corporate and governmental leadership and a willingness to work together.

My expectation is that even though we have already begun, we will recognize that transition in the United States will be well underway by 2030 and largely completed by midcentury. Technological breakthroughs could accelerate this transition. Many of the changes needed require large investments of capital. Modernizing the electric grid is probably the heaviest investment lift, followed by waste sorting and mining. Protecting parkland in the United States will not be a financial burden, but it will require political will and skill.

While the electric grid is complicated and expensive, electric vehicles will become increasingly visible on the American market within a decade. Of course, replacing the planet's two billion motor vehicles will take at least until midcentury—unless, like smart phones, the price and features of the technology accelerate the process. Electric vehicles will be one of the more visible indicators of this transition. I also think that if corporate, institutional, and governmental greenhouse gas targets are reached or nearly reached, we will see many creative approaches to addressing environmental sustainability issues. It is not obvious how these ambitious targets will be reached without new technology and creative organizational adaptation.

It is true that massive federal funding for infrastructure will accelerate the transition. Some of that funding was allocated in the bipartisan infrastructure bill of 2021. But even if the federal government sits on the sidelines, state and local governments and major corporations will move ahead regardless. European nations and China will also move in this direction.

THE IMPORTANCE OF NEW TECHNOLOGY
AND ORGANIZATIONAL INNOVATION

Technology is key. Waste sorting and mining could generate revenues that help recover investments. Better technologies for processing food waste and sewage are needed. Advances in solar cell technologies and battery technologies, along with lower prices, could accelerate decarbonization. If people can charge their cars with low-cost home solar power, then electric vehicle adoption would accelerate dramatically. The pace of technological development is unpredictable, but the obvious need and the potential market provide plenty of motivation. Some new technologies that we will need will never have a private market. The most prominent example is carbon capture and storage. It is clear at this point that the accumulation of greenhouse gases in the atmosphere have already warmed our planet. Someday, we will want to remove those gases and store them safely. In all likelihood, that will be a piece of government infrastructure that will need to be funded by general tax revenues. My guess is that such a move will take place only *after* we have decarbonized our economy and want to reduce the long-term damage caused by the fossil fuel era.

Creativity and organizational innovation are also essential. I've directed Columbia University's Masters in Environmental Science and Policy program since 2002 and its Masters in Sustainability Management program since 2010. Our goal has been to create a cadre of problem-solving sustainability professionals. Programs like ours have proliferated in American universities. Many of our graduates work for large, prominent organizations in the public, private, and nonprofit sectors. All are united in their commitment to achieving the Sustainable Development Goals (SDGs) from the United Nations. But a growing number of these graduates are interested in forming new organizations to address the planet's crisis of environmental sustainability. This has been a long-term trend, particularly in the more private sector–oriented sustainability management program, but I see the tendency growing, and I believe that it represents a powerful and creative force that is exciting and worthy of attention and reflection.

A number of our current students and graduates have worked in start-ups, several have started sustainability consulting firms, and many are working on creative products, services, and technologies in established organizations. The common theme is innovative problem solving. Our sustainability management program includes a course on sustainable entrepreneurship, and a number of other courses in that master's program address issues related to financing and operating start-ups, as well as analyzing risk. Student demand is pushing our curriculum to respond to this hunger to understand the management challenges of starting a new enterprise.

These new organizations and new ideas are particularly important in the field of environmental sustainability because without new and creative thinking, the trend lines of business as usual seem to bring our planet to gloom and doom. In July 2021, in the *Journal of Cleaner Production* Sarah Tiba, Frank J. van Rijinsoever, and Marko P. Hekkert observed:

> Today's sustainability challenges call for the help of radical innovators like those startup founders who are finding creative solutions to pollution, the unsustainable use of resources and the spread of diseases. In fields ranging from clean energy to health treatments, innovative startup firms often translate scientific findings into actionable solutions that can reach a global audience. The new business models established by startup firms are simultaneously breaking open existing arrangements . . . and helping to solve today's pressing societal and environmental challenges. (Tiba et al. 2021)

In sum, we need new and creative technologies, but also new and creative business models and institutional arrangements. Sharing economy start-ups such as Lyft, Rent the Runway, Uber, and Airbnb use smartphone technologies and new forms of revenue generation to invent products that consume fewer resources than traditional business models. Tesla's ability to compete with established car companies was in part due to its freedom from the established thought patterns and standard operating procedures of giant automobile corporations. Tesla's creative advances in battery technology, software design, and rare earth mineral recycling were critical in enabling the company to overcome its lack of production process experience and numerous failures in new product rollouts.

Electric vehicles, renewable energy generation, and greenhouse gas reductions in heavy industries such as cement manufacturing all require creative, nonlinear technical and management innovation. While I would never advocate discarding the production, sales, and distribution capacities of large, established organizations, the energy and creativity of start-ups can both challenge these older organizations and, if acquired by larger businesses, be agents of change.

The technological and managerial challenges of environmental sustainability are profound. I know that some environmental advocates believe that we now have all the technology we need to decarbonize—all that is missing is political will. It is true that some of the resistance to this progress comes from fossil fuel companies and others hoping to recover their sunk costs and seeking to use the political process to block change. However, if renewable energy technology were more accessible, more reliable, and less expensive than fossil fuels, no amount of political manipulation could block the new technology. The fact that renewable energy is less expensive than fossil fuels is not enough. It is more difficult to access. Renewable resource technology is improving daily, but for it to drive fossil fuels and the throwaway economy out of the marketplace, it must dominate existing technologies and business models.

Electric vehicles are among the first tests at scale. They have many advantages over internal combustion engine vehicles. But currently it is easier to find a gas station than a charging station, and electric vehicles are more expensive to purchase, even if they are probably cheaper to operate. Batteries with longer range, faster charging, and lower prices could drive fossil-fueled vehicles from the marketplace, but that will require technical and managerial innovation. Governmental tax subsidies can help for a short time, but for enduring change to occur, the new technology must completely dominate the old.

Environmental sustainability requires government investment in research, infrastructure, finance, and market development, but it also requires a creative, agile, and innovative private sector. While government intervention may be necessary, it is far from sufficient. Renewable energy, energy efficiency, electric vehicles, and resource recovery must prove themselves in the marketplace. The concern that young people have about climate change and the degradation of our natural environment creates favorable conditions for products and services that claim to be greener. Organizations seeking to be green have an easier time attracting talent than those that ignore these issues. Entrepreneurial start-ups focused on sustainability can provide creative ideas and the desperation-driven hard work that is often displayed by hungry, marginal new organizations.

Many educational institutions like the one I work for are doing a great deal to encourage their students to begin new enterprises. We are including start-up education in our management curriculum and at Columbia, resources are being allocated to an initiative called Columbia Entrepreneurship, which is designed to encourage student start-ups. I am seeing similar efforts at many other universities. Start-ups are now part of our professional school curriculum, but as important as they are, they don't always succeed. After a few years, a sustainability start-up has three places to be: (1) larger and better established, (2) bought by a larger and better-established company, or (3) out of business. Regardless of the outcome, those involved in the enterprise learn a great deal, and hopefully even those who fail learn lessons that influence them professionally. I am impressed and amazed by the work that I've seen by my students and graduates as they translate their environmental principles into practical sustainability practices in a world desperate for innovation.

CONCLUSION: TOWARD A BROAD-BASED ENVIRONMENTAL SUSTAINABILITY CONSENSUS

What is truly essential is a broad American (and then global) consensus about the need to transition to an environmentally sustainable economy. In the United States, that means that we need to build a broad, nonideological coalition. The polarization of politics can be focused on issues unrelated to breathing, drinking clean water, eating healthy food, and avoiding catastrophic weather events. Even a politically polarized nation is occasionally capable of a unified response to a true crisis. We saw that in early 2022 during the Russian invasion of Ukraine. We need to transition our economy to environmental sustainability, and the degradation of our environment creates a true crisis. But crisis creates opportunity if we somehow have the wisdom to pursue it.

As I've indicated throughout this book, government plays a key role in accelerating this transition, but the real, on-the-ground action must take place in the private sector. Many business leaders see that there is money to be made in the green economy. The changes in financial markets reflect this understanding. A more crowded planet requires that we pay more attention to our use of natural resources. Recycling must move beyond household do-gooderism to massive investment in waste sorting and mining research and investment. We will mine our garbage only when it is more cost effective than mining the planet.

While the political currents of populism and mindless antiscience remain strong in American society, the people running businesses remain focused on the reality of the markets and the planetary conditions that impose risks to supply chains and operations. Extreme weather due to climate change is not a "belief" in the business world; it is a reality. Political players may have the luxury of creating their own facts, but a broken supply chain or a flooded retail outlet can't be wished away. Human and ecological health are interconnected, and that fact provides the basis for my hope that human ingenuity and our desire to maintain and even improve our lifestyle will keep the creative drive for an environmentally sustainable economy underway.

The transition to environmental sustainability is continuing. It is not perfect, and there have been and will be many setbacks along the way. Crises like COVID-19 and Russia's invasion of Ukraine may seem like interruptions, but paradoxically they have the effect of reinforcing our sense of global interdependence. That awareness is central to resolving the climate crisis and the broader crisis of environmental sustainability. I recognize that some may see the glass that I see as one-quarter full as three-quarters empty. I have been a student of environmental policy for nearly a half-century, and the issue that I've worked on has gone from the periphery of public attention and concern to the center. I am confident that we will make this transition before it is too late. I am confident in the future of humankind.

WORKS CITED

Apple. 2021. "Apple Environmental Progress Report." https://www.apple.com/environment/pdf/Apple_Environmental_Progress_Report_2021.pdf.

Collier, Cassie, Helena Fruscio, Helen Lee, and Janet Tan. 2015. *New York Apparel Cluster*. Cambridge, MA: Harvard University. https://www.isc.hbs.edu/Documents/resources/courses/moc-course-at-harvard/pdf/student-projects/New_York_City_Apparel_Cluster_2015.pdf.

Davenport, Coral. 2016. "As Wind Power Lifts Wyoming's Fortunes, Coal Miners Are Left in the Dust." *New York Times*, June 19, 2016. https://www.nytimes.com/2016/06/20/us/as-wind-power-lifts-wyomings-fortunes-coal-miners-are-left-in-the-dust.html?ref=business&_r=1&mtrref=undefined&gwh=F2AC066020DDFBDF79870BAC51794600&gwt=pay.

Edminster, Ann. 2020. "Zero Energy Residential Buildings Study 2019–2020: Inventory of Residential Projects on the Path to Zero in the U.S. and Canada." EEBA Team Zero. https://www.sips.org/documents/Zero-Energy-Residential-Inventory-2019-2020.pdf.

Jaffe, Amy. 2021. "This Was the Year Investors and Businesses Put Big Bets on Climate." *Wall Street Journal*, December 13, 2021. https://www.wsj.com/articles/investors-climate-2021-11638372735?mod=hp_jr_pos3.

Morese, Nick. 2020. "Nurturing a More Diverse Pipeline of Talent." *Etsy Impact*, January 29, 2020. https://medium.com/etsy-impact/nurturing-a-more-diverse-pipeline-of-talent-ed71d9e08577.

Mozen, Chelsea. 2021. "Etsy's 2021 Ecological Impact Goals." *Etsy Impact*, March 25, 2021. https://medium.com/etsy-impact/etsys-2021-ecological-impact-goals-f81df175a639.

New York Times. 1981. "New York City's Decline in Manufacturing Gained Momentum in 1980," March 22, 1981. https://www.nytimes.com/1981/03/22/nyregion/new-york-city-s-decline-in-manufacturing-gained-momentum-in-1980.html.

New York Times. 2021. "Energy Efficient Isn't Enough, So Homes Go Net Zero," November 16, 2021. https://www.nytimes.com/2021/11/16/business/net-zero-homes.html?searchResultPosition=4.

Peterson, Elizabeth. 2020. "90 Percent of S&P 500 Index Companies Publish Sustainability Reports in 2019, G&A Announces in Its Latest Annual 2020 Flash Report." Governance and Accountability Institute, July 16, 2020. https://www.globenewswire.com/news-release/2020/07/16/2063434/0/en/90-of-S-P-500-Index-Companies-Publish-Sustainability-Reports-in-2019-G-A-Announces-in-its-Latest-Annual-2020-Flash-Report.html.

Tiba, Sarah, Frank J. van Rijnsoever, and Marko P. Hekkert. 2021. "Sustainability Startups and Where to Find Them: Investigating the Share of Sustainability Startups across Entrepreneurial Ecosystems and the Causal Drivers of Differences." *Journal of Cleaner Production* 306 (July): 127054. https://doi.org/10.1016/j.jclepro.2021.127054.

Vanderhelm, Mark. 2021. "Setting Records, Walmart Continues Moving Toward Becoming a Totally Renewable Business." Walmart, April 29, 2021. https://corporate .walmart.com/newsroom/2021/04/29/setting-records-walmart-continues-moving -toward-becoming-a-totally-renewable-business.

Walmart. 2021. "Walmart Creates Industry First by Introducing Science-Based Targets for Supply Chain Finance Program," December 8, 2021. https://corporate.walmart .com/newsroom/2021/12/08/walmart-creates-industry-first-by-introducing -science-based-targets-for-supply-chain-finance-program.

Walmart. 2022. "Climate Change." https://corporate.walmart.com/global-responsibility /sustainability/planet/climate-change.

INDEX

Printed in the USA
CPSIA information can be obtained
at www.ICGtesting.com
LVHW091920041124
795688LV00034B/1045